BIOLUMINESCENT MICROBIAL BIOSENSORS

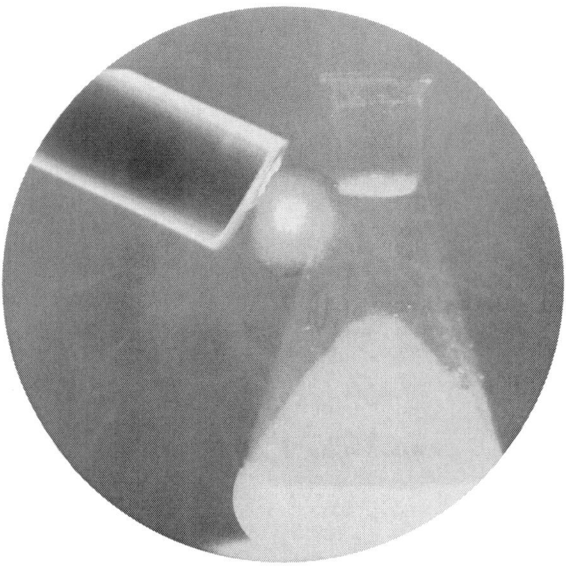

Pan Stanford Series on the High-Tech of Biotechnology

Robert S. Marks
Series Founding Editor

Avram and Stella Goldstein-Goren Department of Biotechnology Engineering
National Institute for Biotechnology Engineering
Ilse Katz Institute for Nanoscale Science & Technology
Ben Gurion University of the Negev
Israel

Titles in the Series

Published

Vol. 1

Nanoantenna: Plasmon-Enhanced Spectroscopies for Biotechnological Applications

Marc Lamy de la Chapelle and Annemarie Pucci, eds.

2013

978-981-4303-61-3 (Hardcover)
978-981-4303-62-0 (eBook)

Vol. 2

Viral Diagnostics: Advances and Applications

Robert S. Marks, Leslie Lobel, and Amadou Alpha Sall, eds.

2015

978-981-4364-43-0 (Hardcover)
978-981-4364-44-7 (eBook)

Vol. 3

Electrochemical Biosensors

Serge Cosnier, ed.

2015

978-981-4411-46-2 (Hardcover)
978-981-4411-47-9 (eBook)

Vol. 4

Nanomaterials for Water Management: Signal Amplification for Biosensing from Nanostructures

Robert S. Marks and Ibrahim Abdulhalim, eds.

2015

978-981-4463-47-8 (Hardcover)
978-981-4463-48-5 (eBook)

Vol. 5

Bioluminescent Microbial Biosensors: Design, Construction, and Implementation

Gérald Thouand and Robert S. Marks, eds.

2015

978-981-4613-65-1 (Hardcover)
978-981-4613-66-8 (eBook)

Forthcoming

Vol. 6

Fibre-Optic Immunosensors and Biosensors

Robert S. Marks, ed.

Pan Stanford Series on the High-Tech of Biotechnology
Volume 5

BIOLUMINESCENT MICROBIAL BIOSENSORS

Design, Construction, and Implementation

edited by
Gérald Thouand
Robert S. Marks

PAN STANFORD PUBLISHING

Published by

Pan Stanford Publishing Pte. Ltd.
Penthouse Level, Suntec Tower 3
8 Temasek Boulevard
Singapore 038988

Email: editorial@panstanford.com
Web: www.panstanford.com

British Library Cataloguing-in-Publication Data
A catalogue record for this book is available from the British Library.

Bioluminescent Microbial Biosensors: Design, Construction, and Implementation

ISBN 978-981-4613-65-1 (Hardcover)
ISBN 978-981-4613-66-8 (eBook)

Printed in the USA

Contents

Preface

Bioluminescence is a phenomenon reported earlier on in history, during antiquity, and as experimental science developed, so did the understanding of the physiological basis of the phenomenon. In our molecular biology age, which coincided with the technical achievements in optoelectronics, it became clear that one could converge these fields to form amazing devices such as luminescent microbial biosensors. Designing a microbial biosensor as a scientific object applied to different areas remains a technical challenge. So our book, in a scant overview of 10 chapters, conveniently divided into three sections going from the bioreporter microorganisms to the measurement platform, brings to account the technical challenges some researchers were faced with. Both the beginner and the experienced researcher interested in the domain will find the book useful, and we hope it will entice them to probe our world, a world both coeditors spent many years contributing to. In our chapters we emphasize the technical concept, the issues involved, and how the proof of concept was reached. The reader will discover how we may use bioluminescent microorganisms to our needs in monitoring the toxicity of water.

Microbial biosensors experienced a new rise since their unique position to sense the toxicity event of chemical compounds in environment, and this new book tries to contribute to this momentum.

Gérald Thouand
Robert S. Marks

Chapter 1

Fully Automated Fluidic Analyzer for Food Quality Assessment Using Bioluminescent Biosensors

Efstratios Komaitis,[a] Efstathios Vasiliou,[a] Dimitrios Georgakopoulos,[b] and Constantinos Georgiou[a]

[a] *Chemistry Laboratory, Agricultural University of Athens, 75 Iera Odos, 11855 Athens, Greece*

[b] *Microbiology Laboratory, Agricultural University of Athens, 75 Iera Odos, 11855 Athens, Greece*

egvasiliou@yahoo.com, dgeorga@aua.gr, cag@aua.gr

1.1 Introduction

The current legislation in the European Union (EU) requires that water quality and the degree of contamination be assessed using chemical methods (European Groundwater Directive). Such methods do not consider the synergistic or antagonistic interactions that may affect the bioavailability and toxicity of pollutants in the environment [1]. Bioassays are methods for assessing the toxic impact of whole samples on the environment and for screening environmental samples before going onto detailed chemical

Bioluminescent Microbial Biosensors: Design, Construction, and Implementation
Edited by Gérald Thouand and Robert S. Marks

ISBN 978-981-4613-65-1 (Hardcover), 978-981-4613-66-8 (eBook)
www.panstanford.com

analyses that can be time consuming and expensive and do not allow monitoring [2]. The utilization of organisms possessing *lux* genes [3] gained significant importance during the last decade since the toxicity bioassays have been recognized as essential tests with chemical analyses [4]. The widely used marine photobacterium *Allivibrio fischeri* is a self-maintained luminescent unit. The level of in vivo luminescence reflects the metabolic rate of luminous bacteria and the integrity of the bacterial cells [5].

In recent years, the development of whole-cell biosensors has found increasing interest due to the capability of whole cells to convert complex substrates using specific metabolic pathways [6] and because potential applications of whole-cell biosensors for monitoring of typical *sum parameters*, such as toxicity [7], biological oxygen demand [8], xenobiotic compounds [9], or heavy metals [10] cannot be monitored using enzyme-based sensors. The opportunity to module metabolic activities of specific cells can additionally be used for drug screening [11] and combinatorial approaches for drug discovery [12]. Additionally, microbial biosensors have been successfully applied for the specific determination of single components such as, for example, glucose, fructose, xylose, and alcohols [13]. A general advantage of microbial biosensors is that living cells are continuously repairing their integrated enzyme activities and enzyme cascades. This is a clear advantage in comparison to biosensors based on labile biological recognition elements (e.g., enzymes). On the other hand, the development of such biosensors allows the use of immobilized organisms that maintain their physiological status and thus the results obtained will represent the natural responses.

A further search of the literature shows that few articles describe the implementation of these devices in automated analyzers for the construction of integrated analytical instrumentation for real sample analysis and monitoring [14]. Furthermore, there are not many applications of whole-cell biosensors in flow techniques. The main reason for this is that when adding whole-cell organisms in solvents, suspensions are produced (instead of homogenous chemical solutions). Suspensions lead to low repeatability experiments, especially when flow analysis is required (liquid chromatography [LC], high-performance liquid chromatography [HPLC], etc.).

This chapter describes a fully automated fluidic analyzer of bioluminescent biosensors for future implementation in food quality assessment through the response to toxic heavy metals [15], antioxidants, and aldehydes using *Allivibrio fischeri* bacteria.

1.2 Analyzer Development

Light detection device: An integrated photomultiplier tube (PMT) incorporating a high-voltage source and divider circuit along with an RS 232 signal output was selected (Hamamatsu HC-135 01), resulting in a compact biosensor device. It should be noted that PMTs are the natural choice when measuring low-level spectral signals.

Flow cell: Due to the use of a PMT detector, a wall-jet flow cell configuration was implemented. Figure 1.1 depicts the design of the prototype based on a previous study by Divritsioti et al. [7]. The assembly was clamped between two stainless steel plates featuring appropriate openings for light detection and passage of the In (entrance of carrier solution) and W (exit of carrier solution) polytrafluoroethylene (PTFE) tubes of 0.8 mm internal diameter.

Temperature control of the flow cell: To maintain immobilized bacteria at their optimal temperature, the flow cell is enclosed in a thermostated aluminium frame using a water bath at 20°C.

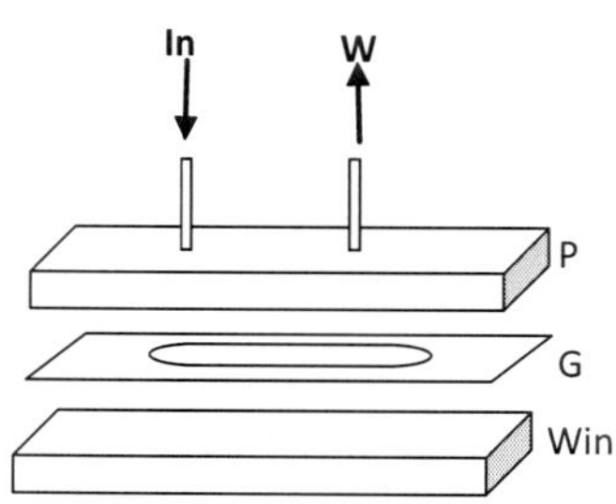

Figure 1.1 Flow cell design. In: carrier solution feed; W: waste; P: Plexiglas plate; G: Gasket, 2 mm thickness, 0.8 cm^2 area, 160 μL volume; Win: optical window from Plexiglas or quartz (PMT detector is opposite). Dimensions after assembly: 2 cm × 1 cm × 1 cm.

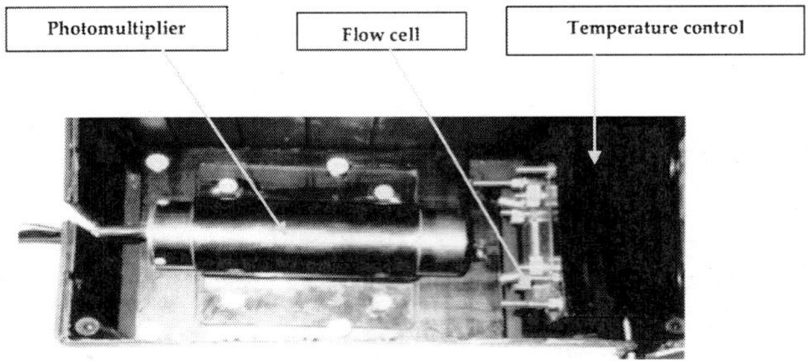

Figure 1.2 Detector unit incorporating a photomultiplier tube, a flow cell, and temperature control.

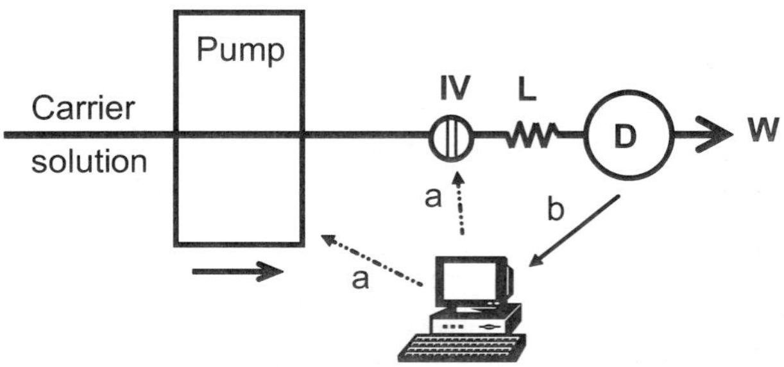

Figure 1.3 Automated flow injection analyzer. D: detector unit; IV: injection valve; L: mixing coil; W: waste; a: digital control signals; b: data acquisition line.

Fluidic analyzer development and optimization: The single line fluidic system that was developed incorporating a detector unit (Fig. 1.2) is depicted in Fig. 1.3. The analyzer design is based on the continuous flow of a carrier solution. The sample loop of the injection valve is loaded with the *Allivibrio fischeri* cell suspension that is subsequently automatically injected in the carrier solution flow. *Allivibrio fischeri* cells are mixed with the carrier solution in the mixing coil and then driven to the detector unit for bioluminescence assessment. The flowchart of the analyzer software is shown in Fig. 1.4.

The percentage inhibition of bioluminescence is assessed with a two-step experimental protocol. The first step is the assessment of *Allivibrio fischeri* bioluminescence by injection in a nontoxic carrier

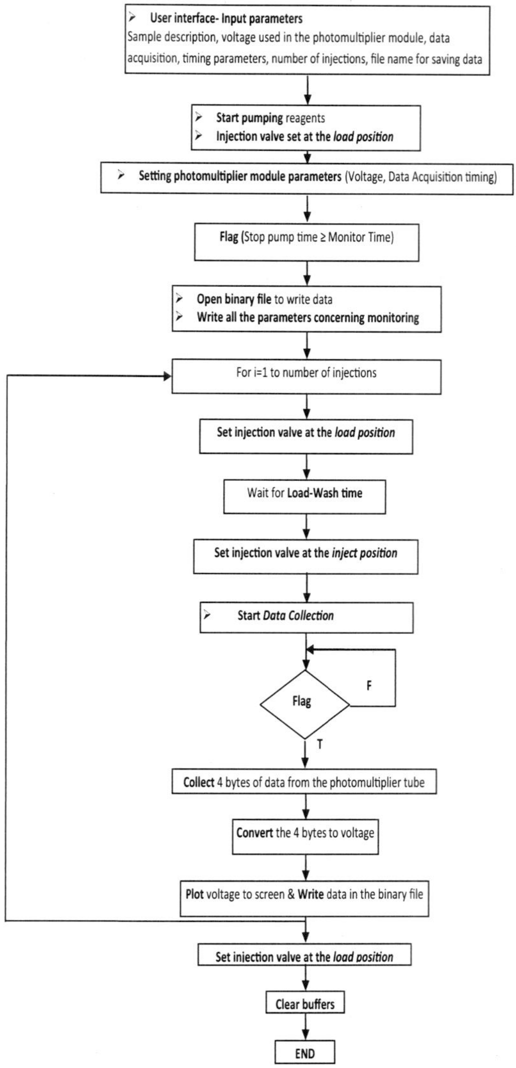

Figure 1.4 Flowchart of the analyzer software.

solution. The second step is the assessment of *Allivibrio fischeri* bioluminescence by injection in a toxic carrier solution (sample).

The percentage bioluminescence inhibition is calculated using the readouts (peaks) from the two injections:

$$\% \text{ Bioluminescence inhibition} = (\text{Peak } 1 - \text{Peak } 2) \times 100/\text{Peak } 1$$

The thus calculated bioluminescence inhibition correlates with the toxic effect of the sample to *Allivibrio fischeri* cells. For evaluation of the peak readouts, both heights and areas have been assessed. *Allivibrio fischeri* washout from the analyzer was a slow procedure, resulting in extensive peak tailing. This was probably due to the shape of the flow cell that does not facilitate the washout of bacteria. To overcome this problem and shorten the analysis time, 2 s after recording the peak maximum, the analyzer pump is set to the maximum flow rate (~10 mL/min). In this way, the turnover time for a single peak was kept to just 40 s. The precision of peak height measurements assessed through 10 injections of the same *Allivibrio fischeri* culture was found to be 0.7% relative standard deviation (RSD). Peak area measurements resulted in lower precision presumably due to the high flow rate used for the washout step.

Suspension of Allivibrio fischeri cells with the fluidic system: *Allivibrio fischeri* strain NRRL-11177 (Dr Lange S.A.) is grown in 20 mL DSMZ No. 6904 broth [4] for 20 h in an orbital incubator at 190 rpm at 24°C [3]. The thus developed stock culture was mixed with 30% glycerine, dispensed in 2 mL vials, and stored at –70°C. Next, 20 μL of the stock culture was inoculated in 50 mL of growth medium and grown overnight. Then the culture was centrifuged and resuspended in artificial seawater, preserving its function but stopping its growth. This suspension was used with the fluidic system.

1.3 Optimization

Three different volumes of the *Allivibrio fischeri* suspension were tried: 80, 100, and 200 μL. The signal increased along the injected volume. Although the signal is maximized when injecting 200 μL of *Allivibrio fischeri* culture, we chose 100 μL in order to minimize consumption of the culture. In this way, 12 injections were feasible using the same culture suspension. Although just 100 μL is injected, another portion of around 400 μL is consumed while filling the injection valve sample loop. This is actually a general disadvantage of the procedure of filling a sample loop through aspiration. Although this disadvantage could be overcome, when using conventional reagents, by replacing aspiration through the use of a syringe, this is not an option when using bacterial suspensions as reagents. Bacterial membranes could be disrupted, resulting in loss of signal. When using an 8 mL culture suspension 16 injections are feasible. It should be noted that *Allivibrio fischeri* culture cannot be used the way reagents are used: Upon standing, due to gravity, bacteria tend to accumulate in the container bottom. This resulted in a gradual signal decrease that was up to 30% for a 1 h operation and diminished precision. To overcome this problem we used continuous stirring of an 8 mL *Allivibrio fischeri* suspension with a magnetic stirrer.

The flow rate was optimized in the range of 0.5 to 2.0 mL/min. At flow rates lower than 1.5 mL/min peaks were not sharp and the whole procedure was slow, resulting in a high turnover time. The flow rate of 1.5 mL/min was chosen as a compromise between an adequate level of signal and at the same time fast washout of *Allivibrio fischeri* cells from the analyzer. Mixing coils of 50, 100, and 150 cm have been tested. The 100 cm mixing coil was selected as a compromise between adequate mixing and analysis time.

1.4 Functional Test of the Fluidic Analyzer

The developed fluidic system was used for the assessment of toxicity of different heavy metals. A typical flow injection toxicity assessment output is shown in Fig. 1.5. It is clear that the system developed was able to assess toxicity due to heavy metals present in water samples. Effective concentrations were in the range of 1.0×10^{-2}M -1.0×10^{-5}M. The fluidic system was applied to aqueous solutions of three heavy metals, namely lead, copper, and mercury. The measurement of inhibition was rapid and the RSD 0.7% ($n = 3$). The detection limits were 10^{-4}M for Pb^{2+}, 10^{-4}M for Cu^{2+}, and 10^{-5}M for Hg^{2+}.

The work presented here is showing the principle of using cells as a reagent in a flow system in order to use the system in water analysis. This experiment proved to be a functional test for the developed fluidic analyzer. Further work is needed to lower

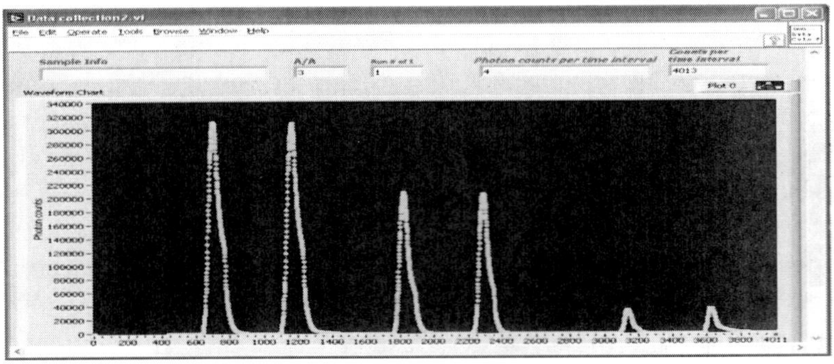

Figure 1.5 Typical toxicity assessment diagram. Horizontal axis: time (s); vertical axis: bioluminescence intensity (PMT counts). Peaks from left to right: blank (1st and 2nd), Cu^{2+} solution 3.0 mM (3rd and 4th) and 10.0 mM (5th and 6th).

detection limits. If needed, detection limits can be modulated by stopping the flow: By stopping the flow for 15 min detection limits can go down to 10^{-7}M, depending on the toxic metal compound.

To measure an *Allivibrio fischeri* dispersion, dilution when injected in the fluidic system, we recorded the signal (D_S) of the bioluminescent suspension using it as a carrier stream and then the signal from 100 μL of the same suspension injected in a carrier stream of deionized water (D_o). The D_S signal divided by D_o gives the dispersion coefficient (dilution), which in the developed fluidic system was calculated and found to be equal to 4.52.

1.5 Determination of Phenolics and Application to Oils

Phenolic antioxidants have significant importance for the food industry and world health. There is a big necessity for their determination. Folin Chiocalteu, 2,2′-azinobis(3-ethylbenzothiazoline-6-sulfonic acid) diammonium salt (ABTS), plus many chromatographic methods (HPLC, gas chromatography mass spectrometry [GCMS], etc.), have been used for quantitative or qualitative analysis of phenolic substances.

In this chapter, a new method using the developed analyzer is described. Optimization led to application of a 3.0 mL/min flow rate

Table 1.1 Percentage inhibition of *Allivibrio fischeri* bioluminescence from phenolics

	1.00×10^{-2} M	1.00×10^{-3} M	1.00×10^{-4} M	1.00×10^{-5} M
Phenol	70.0%	26.2%	15.0%	1.20%
Tyrosol	97.0%	87.0%	51.2%	37.6%
Gallic acid	[a]	[a]	72.0%	56.4%
Ferulic acid	[a]	32.5%	23.0%	8.1%

[a]Not soluble in the 20% methanol aqueous solvent.

and a 100 cm length coil for achieving a precision percentage RSD $<1\%$, $n = 3$.

The most common solvents for extraction of phenolic antioxidants are aquatic solutions of methanol. Thus phenol, tyrosol, gallic acid, and ferulic acid were dissolved in a 20% (v/v) methanol aqueous solution. Concentrations varied from 1.00×10^{-2}M through 1.00×10^{-5}M. Results are presented in Table 1.1.

Tyrosol shows a higher percentage inhibition than phenol. From this initial indicative experiment it can be concluded that the developed system responds to phenolic antioxidants.

In this respect, the response of *Allivibrio fischeri* whole-cell biosensors to extracts of olive oils was tested. The methanol solvent in the carrier stream inhibits *Allivibrio fischeri* bioluminescence almost completely when the carrier stream is pure methanol and 95% and 35% when methanol concentration is 80% and 20% (v/v), respectively. On the basis of these data, although an 80%–20% (v/v) methanol–water mixture is the literature optimum extraction medium, a 20%–80% (v/v) mixture was used. Results are depicted in Table 1.2.

Results of the response of *Allivibrio fischeri* whole-cell biosensors to olive oil extracts are shown in Table 1.2. It is interesting to note that extracts from virgin olive oils exhibit higher percentage inhibition values than those from olive oils. This is presumably due to their higher phenolic content.

To extract phenolics of higher molecular weight we tested aqueous dimethyl sulfoxide (DMSO). The concentration chosen was 5% (v/v) DMSO aquatic solvent. This solvent showed no toxicity to *Allivibrio fischeri* bacteria. Results are depicted in Fig. 1.6.

Table 1.2 *Allivibrio fischeri* whole-cell biosensors response to olive oil extracts

Virgin olive oil	% Inhibition
Mitilinio,[a] Lesvos Agr. Cooperatives	72
Mitilinio,[b] Lesvos Agr. Cooperatives	67
Spitiko,[a] Elaiourgiki	65
Mountainous Regions,[a] Minerva	63
Peza P.D.O,[c] Peza Union	63
Private label,[a] Carrefour	59
Olive oil	
Classic,[a] Minerva	54
Liodoxi,[a] Kore	43

Bottled in [a]plastic transparent,
[b]dark green glass, and
[c]translucent tetra prisma paper.

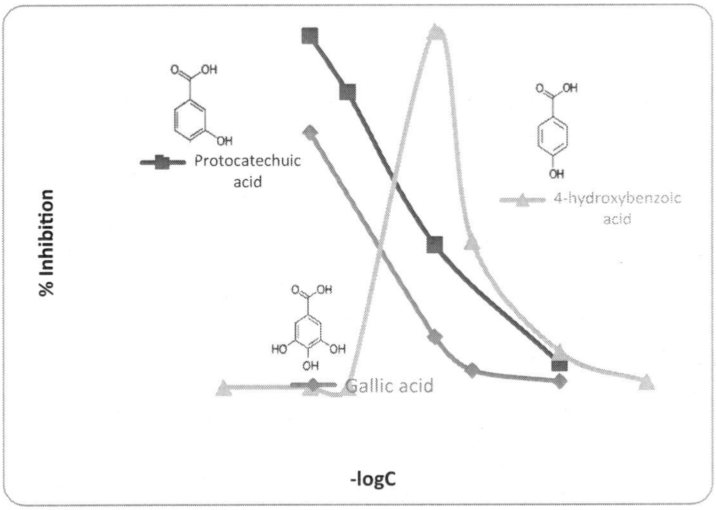

Figure 1.6 Percentage inhibition curves for 4-hydroxy benzoic acid, protocatechuic acid, and gallic acid.

It can be concluded from Fig. 1.6 that *Allivibrio fischeri* bioluminescence is affected by the number of hydroxyl groups of phenolic compounds. These three phenolic compounds differ in structure only in the number of hydroxyl groups. The higher the number of hydroxyl groups, the lower the percentage inhibition observed.

1.6 Determination of Aldehydes

During thermal stressing of oils, significant quantities of aldehydes are produced. Aldehydic compounds such as acrolein [16] and 4-hydroxy-t-2-nonenal [17] are known for their toxicity and have been blamed for causing diseases like cancer and Parkinson. Thus, it was interesting to check the developed analyzer against aldehydes.

In this experiment, 5% DMSO aquatic solutions of butanal, pentanal, and hexanal were tested for their inhibition to *Allivibrio fischeri* bioluminescence. The flow rate applied was equal to 3.0 mL/min. Results are depicted in Fig. 1.7.

It is obvious that the developed analyzer responds to aldehydes. It can also be observed that the percentage inhibition of *Allivibrio fischeri* bioluminescence is proportional to the molecular weight of aldehydes. The higher the molecular weight (hexanal), the higher the

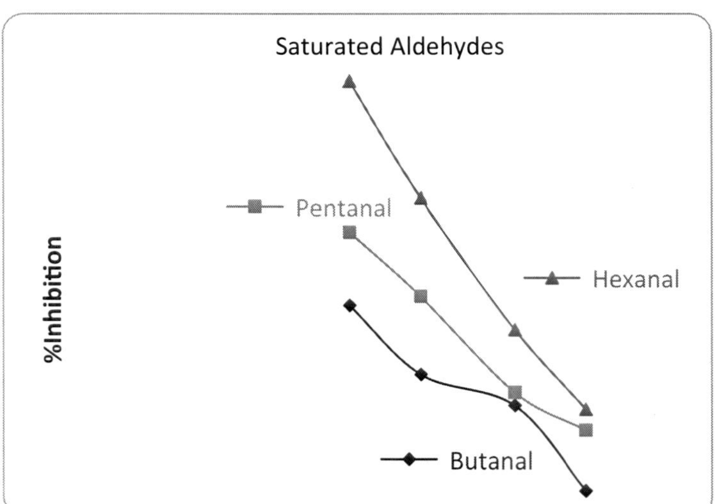

Figure 1.7 Percentage inhibition curves for butanal, pentanal, and hexanal.

percentage inhibition measured. Repeatability in this experiment: % RSD <1%, $n = 3$.

1.7 Conclusions

This chapter describes the development of an automated fluidic analyzer for water toxicity assessment on the basis of *Allivibrio fischeri* bioluminescent bacteria. This fluidic system proved to be useful for assessing toxicity of three heavy metals, depending on their inhibition of *Allivibrio fischeri* bioluminescence. The fluidic system was able to measure inhibition rapidly with a precision of 0.7% RSD. Detection limits were 10^{-4}M for Pb^{2+}, 10^{-4}M for Cu^{2+}, and 10^{-5}M for Hg^{2+}. The biosensor also responded to phenolics, as inhibition decreased along the number of hydroxyl groups. The developed analyzer was also applied on different types of olive oils. Virgin olive oil extracts inhibited *Allivibrio fischeri* bioluminescence more than olive oil extracts. Finally, the fluidic system was able to determine aldehydes; inhibition increased along the molecular weight. This initial work highlights prospects for further application in food quality assessment.

Acknowledgments

This project was cofunded by the European Social Fund and National Resources: O.P. "Education" II. "Environment - Pythagoras II - Funding of Research Groups in Agricultural University of Athens."

References

1. Dewhurst, R. E., Wheeler, J. R., Chummun, K. S., Mather, J. D., Callaghan, A., Crane, M. (2002). The comparison of rapid bioassays for the assessment of urban groundwater quality. *Chemosphere*, **47**, 547–554.
2. Forget, G., Sanchez-Bain, A., Arkhipchuk, V., Beauregard, T., Blaise, C., Castillo, G., Castillo, L. E., Diaz-Baez, M. C., Pica-Granados, Y., Ronco, A., Srivastava, R. C., Dutka, B. J. (2000). Preliminary data of a single-blind,

multicountry trial of six bioassays for water toxicitymonitoring. *Environ. Toxicol.*, **15**, 362–369.

3. Deryabin, D. G., Karimov, I. F. (2009). Simultaneous evaluation of chemiluminescence and bioluminescence in a phagocytic system. *Bull. Exp. Biol. Med.*, **147**, 349–352.
4. Farre, M., Barcelo, D. (2003). Toxicity testing of wastewater and sewage sludge by biosensors, bioassays and chemicalanalysis. *Trac-Trend Anal. Chem.*, **22**, 299–310.
5. El-Alawi, Y. S., Huang, X. D., Dixon, D. G., Greenberg, B. M. (2002). Quantitative structure-activity relationship for the photoinducedtoxicity of polycyclic aromatic hydrocarbons to the luminescent bacteria *Vibrio fischeri*. *Environ. Toxicol. Chem.*, **21**, 2225–2232.
6. D'Souza, S. F. (2001). Microbial biosensors. *Biosens. Bioelectron.*, **16**, 337–353.
7. Divritsioti, M. H., Karalemas, I. D., Georgiou, C. A., Papastathopoulos, D. S. (2003). Flow injection analysis system for l-lysine estimation in foodstuffs using a biosensor based on lysine oxidase immobilization on a gold-poly (m-phenylenediamine) electrode. *Anal. Lett.*, **36**, 1939–1963.
8. Liu, X. M., Germaine, K. J., Ruan, D. (2010). Whole cell fluorescent biosensors for bioavailability and biodegradation of polychlorinated biphenyls. *Sensors*, **10**, 1377–1398.
9. Beyersdorf-Radeck, B., Riedel, K., Karlson, U., Bachmann, T. T., Schmid, R. D. (1998). Screening of xenobiotic compounds degrading microorganisms using biosensor techniques. *Microbiol. Res.*, **153**, 239–245.
10. Petanen, T., Romantschuk, M. (2002). Use of bioluminescent bacterial sensors as an alternative method for measuring heavy metals in soil extracts. *Anal. Chim. Acta*, **456**, 55–61.
11. Liu, Q., Cai, H., Xiao, L. (2007). Embryonic stem cells biosensor and its application in drug analysis and toxin detection. *IEEE Sens. J.*, **7**, 1625–1631.
12. Peters, M., Vaillancourt, F., Xeroux, M. (2010). Comparing label-free biosensors for pharmacological screening with cell-based functional assays. *Assay Drug Dev. Technol.*, **8**, 219–227.
13. Baeumner, A. J. (2003). Biosensors for environmental pollutants and food contaminants. *Anal. Bioanal. Chem.*, **377**, 434–445.
14. Melidis, P., Vaiopoulou, E., Aivasidis, A. (2008). Development and implementation of microbial sensors for efficient process control in wastewater treatment plants. *Bioproc. Biosyst. Eng.*, **31**, 277–282.

15. Komaitis, E., Vasiliou, E., Kremmydas, G., Georgakopoulos, D. G., Georgiou, C. (2010). Development of a fully automated flow injection analyzer implementing bioluminescent biosensors for water toxicity assessment. *Sensors*, **10**, 7089–7098.
16. Seppanen, C. M., Csallany, A. S. (2006). The effect of intermittent and continuous heating of soybean oil at frying temperature on the formation of 4-hydroxy-2-trans-nonenal and other alpha-, beta-unsaturated hydroxyaldehydes. *J. Am. Oil Chem. Soc.*, **83**, 121–127.
17. Miyake, T., Shibamoto, T. (1996). Simultaneous determination of acrolein, malonaldehyde and 4-hydroxy-2-nonenal produced from lipids oxidized with Fenton's reagent. *Food Chem. Toxicol.*, 1996, **34**, 1009–1011.

Chapter 2

Technological Design of Optical Bacterial Biosensors for the Online Environmental Monitoring of Water Pollutants

S. Jouanneau, M. J. Durand, and G. Thouand

UMR CNRS 6144 GEPEA CBAC, De la Courtaisière campus,
IUT de la Roche sur Yon, Biological Engineering Department,
18 Gaston Defferre Bd., 85035 La Roche sur Yon, France
gerald.thouand@univ-nantes.fr

2.1 Introduction

Biosensors developed over the last 10 years mainly touch on 3 business sectors: environmental monitoring, food processing, and the medical industry. Their user-friendliness, moderate cost, and short response time are interesting arguments in favor of these biological tools. Consequently, these biological methods seem to be attractive devices for simultaneous monitoring of both the bioavailable fraction of pollutants and their toxicity level.

These devices are generally dedicated to monitoring online the environmental waters as opposed to off-line systems. An online system is defined as a system wherein the samples are automatically

Bioluminescent Microbial Biosensors: Design, Construction, and Implementation
Edited by Gérald Thouand and Robert S. Marks

ISBN 978-981-4613-65-1 (Hardcover), 978-981-4613-66-8 (eBook)
www.panstanford.com

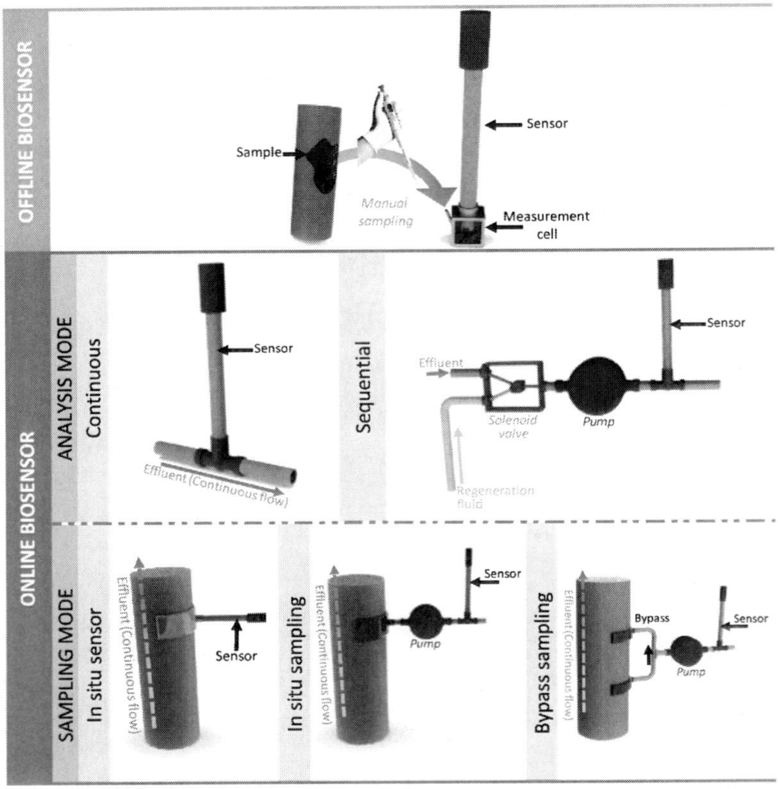

Figure 2.1 Classification mode of bacterial biosensors.

loaded, contrary to the off-line system, which requires a manual injection of the samples [1].

Furthermore, online biosensors are also classified according to their analysis stage (continuous or sequential) and their sampling mode (in situ sensor, in situ sampling, or bypass sampling) (Fig. 2.1) [2]. In a system of continuous analysis, bacteria are exposed continually to the effluent, while with a sequential (or semicontinuous) biosensor, bacteria alternate the exposure and the washing stage. It is also important to define different sampling modes. An in situ sensor is directly positioned into the analyzed site, contrary to the two other sampling modes. With an in situ sampling, the samples are taken in the effluent via a pump and transferred to the measurement cell. With bypass sampling, the samples are taken into a sampling module. For example, it is the case of biosensors using a pretreatment system by tangential flow filtration.

The scientific literature displays two major groups of bacterial biosensors according to the configuration of bacteria either in the liquid phase or the immobilized phase [3–9].

2.1.1 *Liquid-Phase Optical Microbial Biosensors*

For the first category, several biosensors are commercialized: TOXcontrol (microLan), ABEL-TOX^{TM} (Knight Scientific Ltd.), and ToxAlert (UniBest Group Corp.). These continuous systems allow for online monitoring of toxicity in water, thanks to the bioluminescence levels emitted by bioluminescent strains, for example, *Alivibrio fischeri*.

Besides this system, we can quote the experimental systems developed by Gu and Gil [10], Horry et al. [11], and Lee and Gu [12]. The heart of these systems relies on a bioreactor (or several), which allows for the culture and preservation of bacteria in a stable state (exponential growth phase). Nevertheless, the mode of contact between samples and bacteria differs according to the authors. For Gu and Lee, this contact is performed in a secondary reactor. This arrangement limits the disturbances of the bacterial cultures during the addition of the pollutant in the bioreactor. For the system of Horry et al. [11, 13], contact is made directly inside the culture bioreactor, allowing the system to be considerably simplified.

In laboratory conditions, these systems allowed for control of the main parameters influencing bioluminescence (oxygen, temperature, pH, and cellular density). Furthermore, the liquid matrix allows for very fast diffusion of all the compounds (oxygen, nutrient, pollutant) in the bioreactor. The main observed limits concern the biofilm appearance in the bioreactor, disrupting the bioluminescence measurement and the bacteria maintenance mode (only one strain per reactor), which limits the number of analyses, except in the case of the complex system developed by Gu and Gil [10].

Moreover, these biosensors present a common technological constraint linked to the exposure between samples and bacteria.

Indeed, the analysis requires a direct contact between the analyzed sample and the bacteria by injection in the biological reactor. This involves dilution of the sample and consequently a loss of sensitivity [10, 11].

Finally, the maintenance of these biological systems requires deep microbiological knowledge, excluding its use by unskilled staff.

2.1.2 *Immobilized-Phase Optical Microbial Biosensors*

In the second category of biosensors, the bacterial cells are immobilized on a solid support. For that purpose, the bacteria immobilization method does not have to interfere both on the cell activity and on the bioluminescence propagation throughout the matrix. Several methods were developed to immobilize bacteria according to a physical or chemical immobilization [5, 14, 15].

The first physical technique consists of absorption of bacteria on a porous support [16] like a disc of dimethyl silicone [17]. In that case, the disc is introduced into the culture medium to be colonized naturally by the bacteria: A direct connection is formed between the cells and the support (cellulose, glass, cotton, synthetic membrane), thanks to weak bonds (hydrogen bond, van der Waals interaction). In that case, the bonds do not allow the bacteria to be properly fixed to the support.

With the second physical technique, the cells are trapped in a solid matrix of polymer hydrogel. For that purpose, bacteria and polymers are mixed before being polymerized chemically or physically depending on the nature of the polymer (k-carrageenan, agar, agarose, alginate, polyurethane-polycarbomyl sulfonate [PCS], polyacrylamide, poly[vinyl alcohol] [PVA], sol-gel) [14, 18–23]. Bacteria are trapped in the network formed by the polymer [24].

Finally, the last way is to chemically bond the cells to the surface of a solid support [15, 25]. To support grafting, the cells are connected indirectly to the support with an intermediate element, allowing for better fixation (proteins, polymers, antibodies). The strong stability of these chemical bonds excludes the reuse of the immobilization supports.

Contrary to biosensors in the liquid phase, these devices allow for the use of several reporter strains with different kinds of specificities [14, 18, 26]. Furthermore, bacterial immobilization allows for direct contact between the sample and bacteria without any dilution; however, contrary to liquid-based biosensors, these immobilized biosensors lack sensors to control the parameters influencing bioluminescence [27].

In every case, the lack of characterization and evaluation of biosensor performances slows down scientific recognition of these new biological detection tools and their industrial development considerably. Indeed, studies performed in this field generally stayed at the technical and technological development phases, and the developed systems were seldom assessed in real conditions (environmental monitoring, industrial management) for long periods of time.

2.2 Laboratory-Designed Biosensors from the UMR GEPEA

For the past decade, the main topic of our research laboratory was the development of new innovative tools dedicated to the improvement of the environmental monitoring of pollutants. The heart of all these systems relies on bioluminescent bacterial strains (described later) capable of emitting specific bioluminescence in the presence of certain pollutants. Several generations of biosensors, shortly described hereafter, were developed before the design of the last one: Lumisens IV.

2.2.1 *Lumisens I: First-Generation Biosensor*

Lumisens I belongs to the first generation of bacterial biosensors dedicated to environmental monitoring [11, 13, 28]. This biological system described in patent no. FR2831182 is a liquid-phase biosensor (Fig. 2.2). Bacteria are kept in a 100 mL bioreactor under continuous cultivation that is controlled and regulated with several miniprobes (pH, dissolved oxygen and temperature). The measurement of the luminous intensity emitted by a laser diode allowed for the control of cell density in the bioreactor. Bioluminescence levels produced by the bacteria are measured with

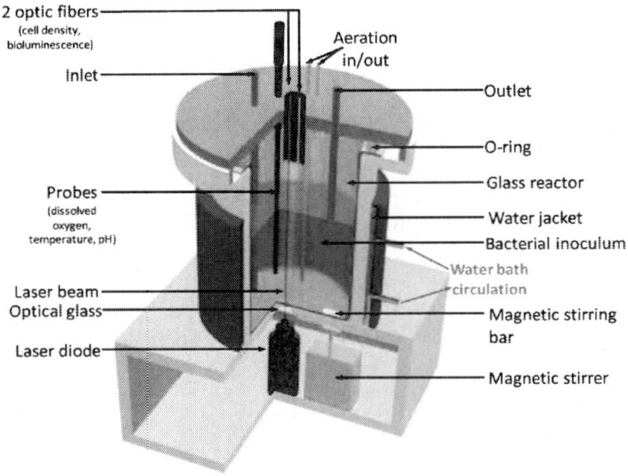

Figure 2.2 3D representation of Lumisens I [11, 13, 28]. The inlet and outlet of the culture media are connected to a peristaltic pump, allowing for continuous culture and sample addition.

a photon-counting photomultiplier. The bioreactor configuration of Lumisens I allows for the integration of only one bacterial sensor and consequently for the analysis of only one compound.

2.2.2 *Lumisens II: Second-Generation Biosensor*

Lumisens II is the first representative of these second-generation biosensors designed in 2007 by Horry et al. [27] and protected by patent no. 2889591 (Fig. 2.3). This second generation of bacterial bioluminescent biosensors is based on bacteria immobilized in a solid matrix of agarose hydrogel. Bacteria are entrapped inside the disposable macrochip in a solid matrix of agarose at 2% (w/w) with a final cell density of nearly 10^8 cells/mL.

This polymer was chosen because of its physical (mechanical resistance, adsorption, and diffusion coefficients), optical (transparent), and chemical (stability in environmental conditions) properties compatible with our specifications [15, 29].

Lumisens II is divided into three physical parts: the fluidic layout, a disposable card, and optical measurement equipment. The first part is dedicated to the circulation and selection of the liquid

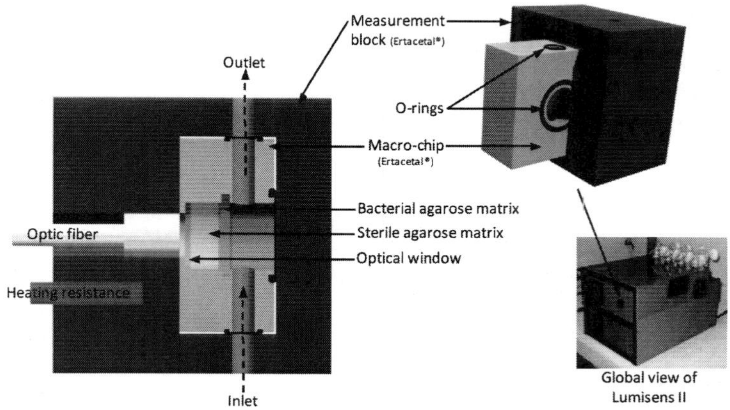

Figure 2.3 Schematic representation of the analytical block of the biosensor Lumisens II [27, 30].

(nutrients or analyzed samples) inside the macrochip controlled with a peristaltic pump (flow rate: 2 mL/min) and a 10-position solenoid valve, respectively. The disposable macrochip, thermally regulated by a heating resistance, is the support of the bacterial sensor, allowing for the exposure of the immobilized bacteria to the circulating fluid and consequently for the detection of the presence of some pollutants in the analyzed samples. This card was designed to use only one bacteria strain like Lumisens I. The last part of the system is dedicated to the measurement of the bioluminescence signal emitted by the immobilized bacteria and transmitted via an optical fiber to a photon-counting photomultiplier.

A solenoid valve and a peristaltic pump allow for the selection of the liquid media exposed to the immobilized bacteria. The bacterial bioluminescence is transmitted to a photon-counting photomultiplier via an optical fiber. The heating resistance allows for the temperature of the immobilized bacteria to be regulated.

2.2.3 *Lumisens III: Third-Generation Biosensor*

Lumisens III [3] also belongs to the second generation of biosensors based on immobilized bacteria. This device is an upgraded version of Lumisens II similar both in the mode of bacteria immobilization

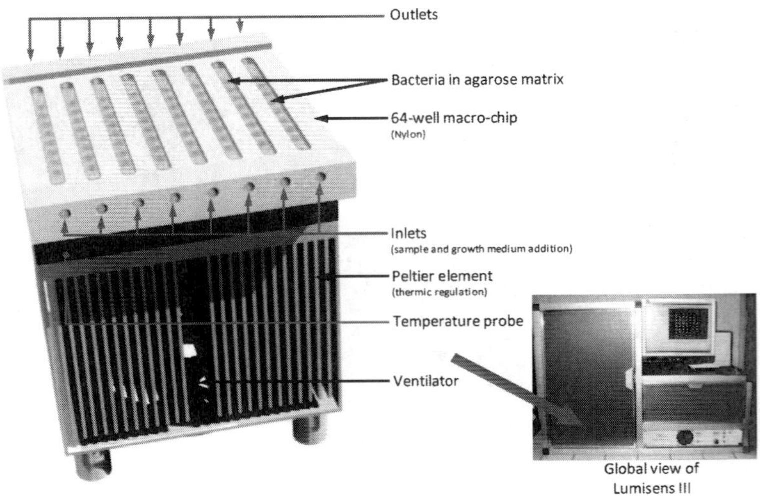

Figure 2.4 3D representation of the Lumisens III measurement block.

in the agarose matrix and the circulation and selection of liquids controlled with a multichannel peristaltic pump (flow rate by channel: 2 mL/min) and a 10-position solenoid valve.

Nevertheless, the design of the disposable macrochip, the measurement method, and the temperature regulation are different (Fig. 2.4). The macrochip features 64 wells split into 8 channels, allowing for simultaneous use of 8 different bacterial strains (one strain per channel). In this configuration, the bioluminescence measurement with a photon-counter photomultiplier proved to be complicated in Lumisens III because of the multitude of wells. Consequently, a charge-coupled device (CCD) camera was adopted, allowing for simultaneous measurement of bacterial bioluminescence in all wells.

The macrochip is divided into eight lines of eight wells each. An electrostatic transparent film is stuck on the surface of the macrochip to ensure its waterproofness.

To improve the temperature regulation, the heating resistance of Lumisens II was replaced by a Peltier element, allowing for the temperature of the macrochip to be increased or decreased depending on the environmental conditions and, on the other hand, to obtain a homogeneous temperature regulation between the 64 wells.

2.2.4 *Advantages and Disadvantages of Three Bacterial Biosensors*

With the probes set, Lumisens I depicts an excellent control of the bacterial biomass inside the bioreactor. Moreover, the liquid matrix containing bacteria allows for fast diffusion of all components (oxygen, nutrients, and pollutants) within the bioreactor. Consequently, the response delay was shorter than the second-generation biosensors: 40 min versus 70 min with the immobilized-phase bacterial biosensors in the same condition. Nevertheless, a biofilm appears on the internal surface of the reactor because of the bacterial growth, disrupting the bioluminescence measurement and the determination of pollution levels. On the other hand, the analysis procedure in Lumisens I entails direct contact between the sample and microorganisms, involving a dilution of the sample and a risk of contamination for the bioreactor (with environmental microflora). Moreover, control of this system requires deep microbiological knowledge, excluding use by a layperson.

By comparison, Lumisens II and III seem more appropriate for daily use because of their disposable cards (simplified maintenance). Nevertheless, the results obtained in previous works showed that these systems were strongly influenced by the environmental microflora, leading to an uncontrollable biofilm (results not published in the present document). Consequently, these devices were not reliable tools for environmental monitoring.

2.3 Lumisens IV: Fourth-Generation Biosensor

This project aims to solve the problem linked to the reliable online monitoring of water pollutants in environmental or industrial contexts. For that purpose, this analytical tool had to meet several use specifications: autonomy, signal stability compatibility during long analysis periods (10 consecutive days), and user-friendliness (without particular knowledge). Given that the first-, second-, and third-generation biosensors were unable to meet these specifications, a new biosensor had to be designed.

This new system (Lumisens IV) is a compromise between the concepts of biosensors and bioassays. It was designed to allow for the detection and quantification of heavy metals in the environment and is based on inducible bioluminescent bacteria, which are maintained in a stable state of preservation after freeze-drying in commercial 96-well microplates. Nevertheless, this system can also be used with other bioluminescent bacterial strains for the detection of other pollutants.

This innovative device is described in detail in the following section.

2.4 Description of the Biosensor Lumisens IV for Heavy Metal Detection

2.4.1 *Bacterial Strains*

The five strains used in this biosensor are bioluminescent engineered bacteria. Four strains, *Escherichia coli* pBZntlux (*luxCDABE*$_{\text{Aliivibrio fischeri}}$ under control of promoter ZntA_P), pBMerlux (*luxCDABE*$_{\text{Aliivibrio fischeri}}$ under control of promoter MerR_P), pBCoplux (*luxCDABE*$_{\text{Aliivibrio fischeri}}$ under control of promoter CopA_P), and pBArlsux (*luxCDABE*$_{\text{Aliivibrio fischeri}}$ under control of promoter ArsR_P), were developed to induce bioluminescence, specifically in the presence of certain metals, and the last one, *E. coli* pBTaclux (*luxCDABE*$_{\text{Aliivibrio fischeri}}$ under the control of promoter Tac_P), is a constitutive strain used to control the toxicity [3, 31].

All strains were cultivated in an acetate medium [31] containing 2.835 g/L CH_3COONa (Panreac), 0.1919 g NH_4Cl (Merck), 0.028 g K_2HPO_4 (Merck), 5 g NaCl (Merck), 0.5 g yeast extract (Merck), 0.1 g tryptone (Biokar Diagnostics), and 50 μg/mL ampicillin sodium salt (Sigma Aldrich).

The strains were engineered according to the same model: a host strain (*E. coli*) was transformed with a plasmid containing bioluminescence genes (*luxCDABE* from *Aliivibrio fisheri*) under the control of an inducible or constitutive promoter. The plasmidic construction of *E. coli* pBMerlux is represented in Fig. 2.5.

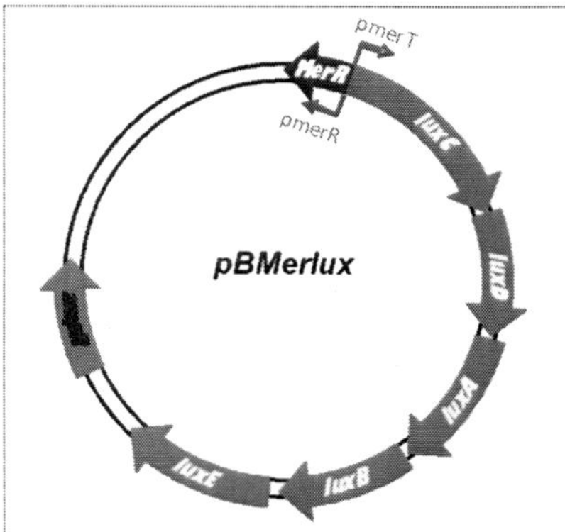

pmerT: inducible promoter
pmerR, MerR: regulating complex (promoter, gene)

Figure 2.5 Plasmidic construction used to transform the host strain.

Before designing a new biosensor, the bioelements had to be characterized to know what the relation was between bioluminescent emission and the physiological state.

The relation between bioluminescence and the growth phase is presented in Fig. 2.6. With the constitutive strain, *E. coli* pBTaclux, the maximum bioluminescence emission was observed during the exponential growth phase after a 2 h culture.

Contrary to *E. coli* pBTaclux, *E. coli* pBCoplux produced the maximum bioluminescent level during the growth stationary phase. The bioluminescence produced by *E. coli* pBMerlux seems to decrease at the end of the exponential growth phase. For other strains, bioluminescence induction is less impacted by the growth phase.

Induction ratio: Bioluminescence emitted by the sample with regard to the control. For inducible strains, the results presented correspond to the bioluminescence levels measured after 60 min of contact with the inducer: cadmium (25 μM), arsenic (25 μM), mercury (2.5 μM), and copper (2.5 mM), respectively, for *E. coli* pBZntlux, pBArslux, pBMerlux, and pBCoplux.

2.4.2 *Bacteria Preservation*

In this project, the freeze-drying method was chosen to preserve cells during long periods of time. This process consists of totally

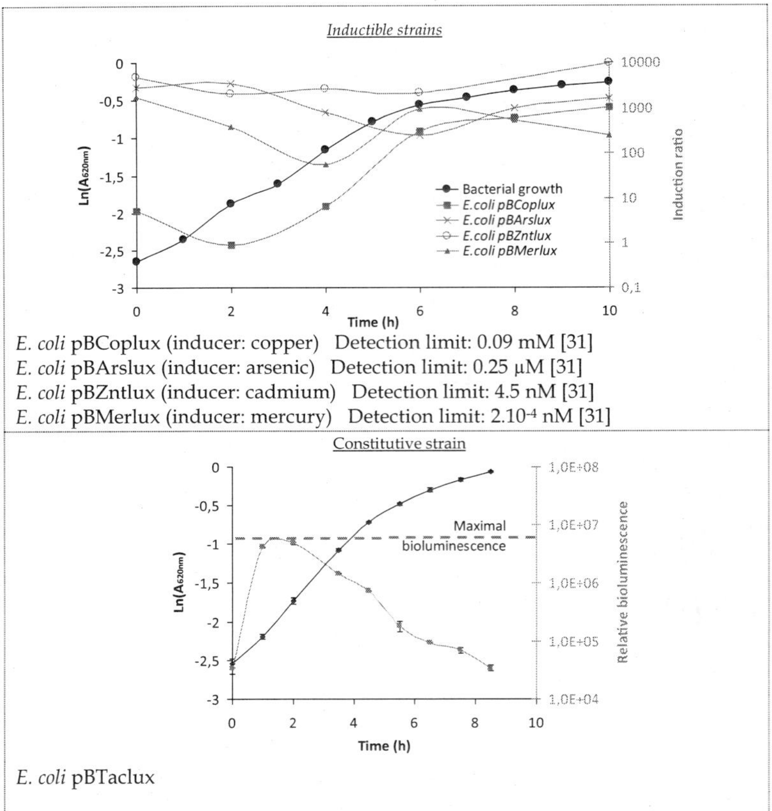

Figure 2.6 Profile of bioluminescence induction according to the bacterial growth phase.

dehydrating the cells by sublimating the cytoplasmic water using a specific pressure (0.05 mbar) at −50°C. The protocol was adapted from Wagner et al. [32] and was detailed by Jouanneau et al. [31]. For this purpose, bacteria are freeze-dried in 96-well microplates with transparent bottoms (56008, Thermo Scientific, Nunc).

This process allows for bacteria to be preserved for at least a month at −20°C, as depicted in Fig. 2.7. From these data, the freeze-dried bacteria harbored the same sensitivity to metal (detection limit: 0.05 μM) and the same induction levels for the entire duration.

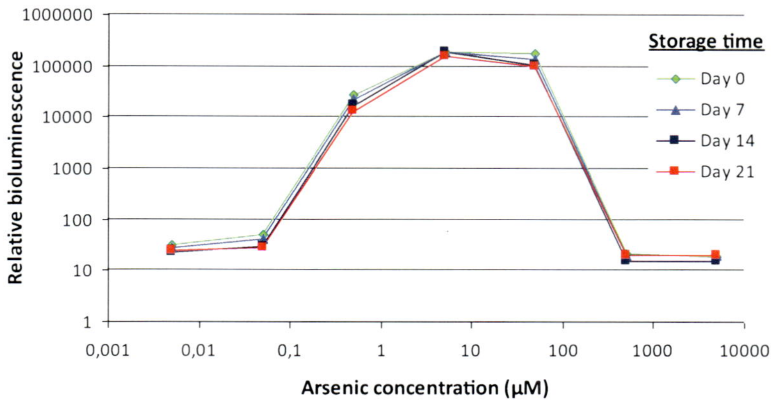

Figure 2.7 Effect of the storage duration on the bacterial sensitivity in the detection of arsenic with *E. coli* pBArslux.

Nevertheless, this storage mode involves a preliminary bacteria rehydration step (100 µL of distilled water is injected in each microplate well containing bacteria) just before the analyses. Moreover, microplates with freeze-dried bacteria are hermetically sealed with an aluminum self-adhesive film (AB0626, Thermo Scientific ABgene) and stored at −20°C.

2.5 Technological Description of Lumisens IV

The first essential step involving computer-aided design and drafting was performed using Solid Edge V.10. Lumisens IV features a set of three technical compartments allowing for liquid circulation, temperature regulation, and bioluminescence measurement [33]. Nevertheless, the main innovation of this new biosensor resides in this data analysis system of biological data, shortly described hereafter.

2.5.1 *Liquid Circulation*

The heart of this compartment is based on a robotized distributor Micro10x^{TM} (Hudson Robotics, Inc., USA). This automation provides a combination of fast filling speeds, high accuracy, and extreme

Figure 2.8 Hudson's Micro10x™ robotic benchtop dispenser.

flexibility. First, its positive-displacement pump head is adjustable in increments of 10 nL and never requires recalibration. Second, Hudson's Micro10x™ allows for the simultaneously distribution of microvolumes in the eight wells of one column of a commercial microplate with accuracy, and third, the linkage arm allows for the penetration of the film, which seals the microplate containing freeze-dried bacteria (Fig. 2.8).

Furthermore, the benchtop dispenser is placed after a solenoid valve (VICI Valco, Interchim), as in Fig. 2.8. This second computer-controlled element is responsible for the selection of injected liquids (distilled water needed for the rehydration of freeze-dried bacteria or analyzed samples). All liquids circulate in a Tygon® pipes (R-3603, internal Ø: 1.6 mm; external Ø: 3.2 mm).

A peristaltic pump (0203-145, TL 150, Medorex) was used to extract wastewater from the system. This pump can also be used to take samples in a dynamic effluent (river, pipe, etc.), as presented in Fig. 2.9.

2.5.2 *Design of the Thermoregulation System*

Bacterial bioluminescence being strongly influenced by the environmental temperature, it was necessary to regulate the temperature inside the measurement chamber. Consequently, the wall of this room was isolated with a flexible film of insulating wool (13 mm, Superwool™ 607™), and the internal atmosphere was maintained at 30°C with a heating device designed in our laboratory with

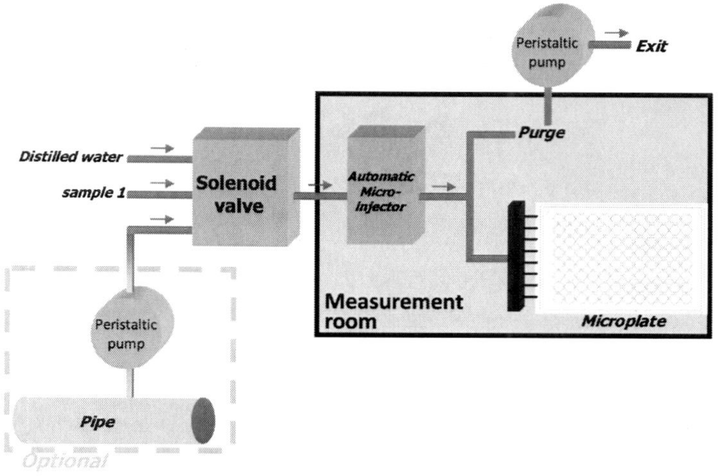

Figure 2.9 Schematic representation of the Lumisens IV fluidic network.

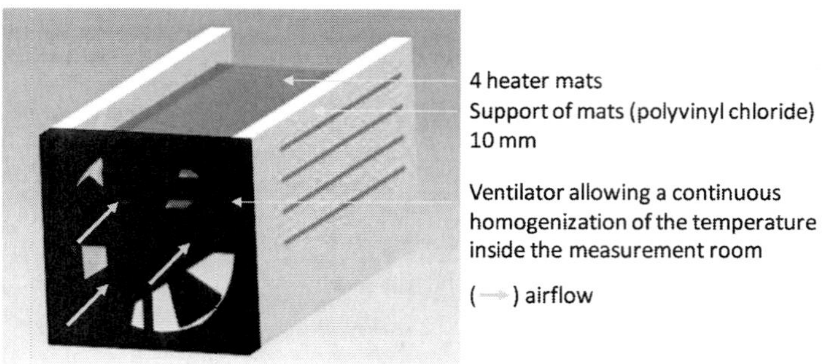

Figure 2.10 Internal heating device used in the Lumisens IV measurement chamber (Solid Edge V.10).

computer-aided design software: Solid Edge V.10 (Fig. 2.10). This device featured four heater mats (100 mm × 150 mm, 15 watts, 731-350, Radiospares), a ventilator (73.1 m^3/h, 541-5075, Radiospares), a temperature probe pt100 (K1P-E-2LS-70, Electrotherm) placed near the microplate, and a temperature controller (reference no. 95/00411579, Jumo).

Figure 2.11 Capture of the bioluminescence signal emitted by bacteria in the biosensor Lumisens IV.

2.5.3 *Bioluminescence Signal Measurement*

Several optical sensors are available to capture the bioluminescence signal emitted by bacterial cells: photomultiplier, complementary metal-oxide semiconductor (CMOS) sensor, and CCD camera. The latter was chosen as a good compromise between sensitivity and simplicity and allowed for simultaneous measurement of the bioluminescence level produced by each of the microplates' bacterial well.

A CCD camera (1N18, Diagnostic Instruments Inc., INSIGHT™) equipped with a 30 cm focal length lens and placed under the 96-well microplate with a transparent bottom (Fig. 2.11) monitored the bioluminescence of the induced freeze-dried bacteria in a totally light-tight measurement room. For that, the walls were insulated with a layer of wool sandwiched between two layers of black expanded polyvinylchloride (4 mm, Forex®).

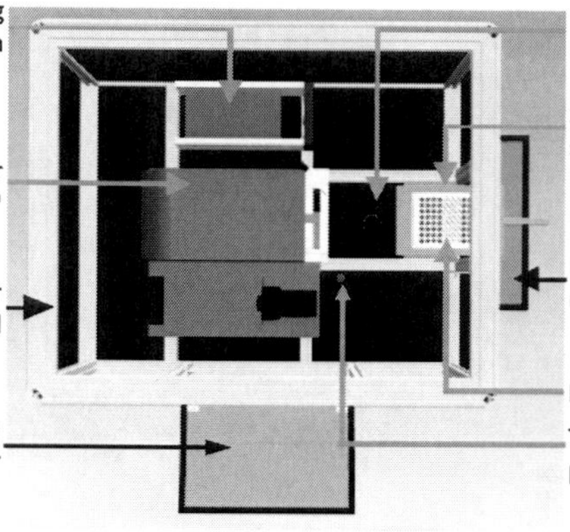

Figure 2.12 Organization of the measurement compartment located in the upper part of Lumisens IV.

2.5.4 *Organization of Lumisens IV*

The measurement compartment is located in the upper part of the biosensor Lumisens IV (Fig. 2.12 and Fig. 2.13). This light-tight compartment dedicated to bacterial analysis contains the heating system and the control probe (pT100), as described above (Section 2.5.2.), the benchtop dispenser, and the camera objective.

Figure 2.13 Global 3D view of Lumisens IV.

The dispenser is connected to the solenoid valve located in one of the lower compartments with a pipe (internal Ø: 1.6 mm, Tygon® R-3603). This compartment allows withdrawing the rising liquids used to wash the dispenser between two measurements.

The last compartment is dedicated to the manual control of the biosensor: temperature instruction in the measurement room (30°C ± 1°C), injected liquids, and Micro10x™ parameters.

2.6 Computer Interface of Biosensor Control and Data Processing

2.6.1 *Control Software*

There are three computer-controlled elements: the CCD camera, the solenoid valve, and the benchtop dispenser. Micro10x™ pictures of the bacterial bioluminescence taken with the CCD camera were put on SPOT™ V.4.6.4.2. (Diagnostic Instruments Inc.). This computer tool allows for the control of the integration time, sensor sensibility, and other parameters. Before every analysis, several pictures without bacteria were executed to estimate the biosensor background and subtract it automatically from the following pictures with bacteria. The pictures are provided in Tagged Image File Format (TIFF).

The solenoid valve is controlled with a communication utility from Microsoft Windows, namely HyperTerminal. This software allows for liquid injected in the microplate wells to be selected in real time or a sequence of successive selections of different liquids (distilled water or analyzed samples) to be edited before the well injection with the benchtop dispenser Micro10x™.

The last control software is dedicated to the dispenser SoftLinx™ (Hudson Robotics, Inc.). It is a powerful multitasking application used to plan and run lab automation work cells. This computer tool was used to edit the sequence of operations performed on the microplate containing the freeze-dried bacteria. For one analysis (1 column of the 96-well microplate), the dispenser performed the following operation sequence:

(1) Rinsing of injectors with distilled water

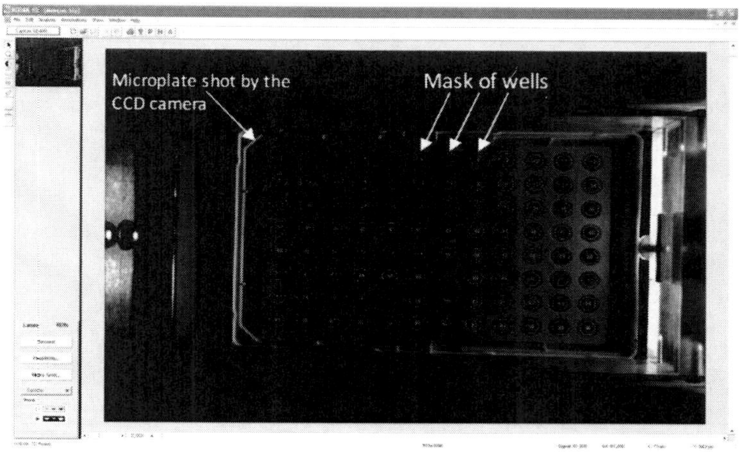

Figure 2.14 Mask of wells preformed by Kodak1D 3.6 to quantify the bioluminescence produced by the freeze-dried bacteria inside the microplate.

(2) Perforation of the aluminium self-adhesive film
(3) Injection of 100 μL of distilled water (bacteria rehydration)
(4) Waiting time of 30 min (rehydration delay)
(5) Rinsing of injectors with the analyzed sample
(6) Injection of 25 μL of the analyzed sample

This sequence was repeated for every column of the microplates.

2.6.2 *Data Processing*

2.6.2.1 Picture processing

A picture processing step is necessary to convert the pictures taken by the biosensor with the CCD camera via SPOT^{TM} (described above) into bioluminescent intensities. For that, we used Kodak1D 3.6 (Kodak, Scientific Imaging System). A mask was placed on the pictures (Fig. 2.14), allowing us to recognize each well. In the second step, the software determines the bioluminescence level (numeric value) of every well, individually, from the sum of the pixel intensities.

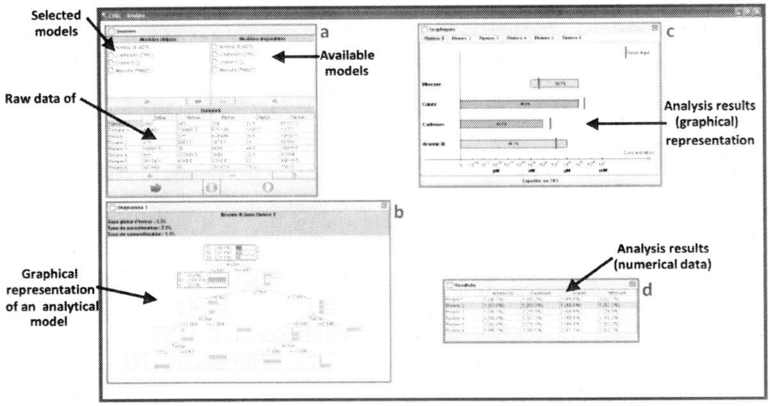

Figure 2.15 Screen capture of Metalsoft's interface dedicated to bioluminescence data analysis.

2.6.2.2 Data analysis

To improve the data analysis process, we developed an analytic software dedicated to the interpretation of the bioluminescence data provided by the bacterial set composed of five bioluminescent bacterial strains described above (Section 2.4.1). This software is described in a previous publication [31].

This software uses the bioluminescence data of the bacterial set to effectively predict the concentration level of four metals (cadmium, arsenic, copper, and mercury) in a mixture and environmental samples using several statistical models (Fig. 2.15).

Metalsoft's interface was developed to be user friendly. The first window (Fig. 2.15a) allows for the selection of the identification models and the input of the analyzed bioluminescence data provided by the bacterial set of Lumisens IV. After the computer processing, three other windows are available. The window presented in Fig. 2.15b represents graphically the classification algorithm, allowing for the estimation of the metal concentrations in the analyzed samples using the bioluminescence data. And the two last windows on the right (Fig. 2.15c and Fig. 2.15d) show the estimated metal concentrations in the samples from both the selected models and bioluminescence data.

2.7 Environmental Application of the Biosensor Lumisens IV

For the validation of our bacterial biosensor, we used environmental samples (water from the outlet of a wastewater treatment plant) that were artificially contaminated with metals. For this last step, we programmed the biosensor Lumisens IV to perform one daily analysis cycle during 1 week. The pictures provided by SPOT^{TM} were transferred to Kodak1D to be analyzed. In parallel, each analysis cycle was completed with a chemical quantification of mercury, cadmium, arsenic, and copper by traditional methods to check the results of Lumisens IV.

The results displayed in Fig. 2.16 show a good correlation between the metal quantification by the biosensor method and by the chemical methods. Indeed, during the 7-day analysis period, the chemical values are included in the concentration intervals estimated by Lumisens IV. For example, the biosensor analysis performed during the second day allowed for the detection of a pollution event with cadmium. Moreover, the biosensor predicted also contamination in this sample with mercury and copper (result not shown). The concentration intervals estimated by Metalsoft using the bioluminescent data of the bacterial set were [25 nM; 250 nM], [0.5 nM; 5 μM], and [5 μM; 50 μM], respectively, and the concentrations measured with the chemical methods were 142.5 nM, 109.9 nM, and 49.52 μM, respectively (result not shown, except for cadmium). The chemical analysis confirmed the arsenic-

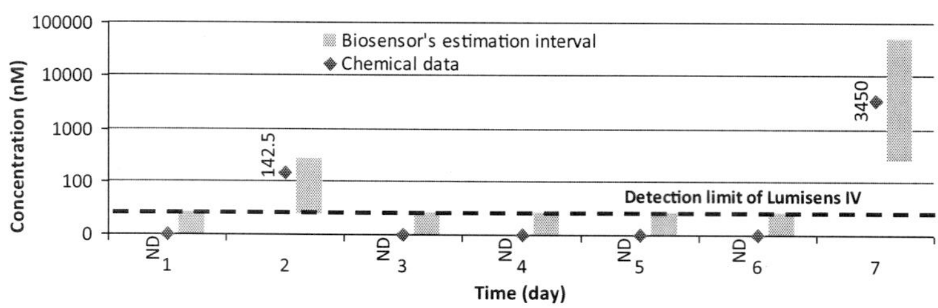

Figure 2.16 Comparison between the biosensor and chemical quantification for cadmium (samples: water from a wastewater treatment plant outlet). ND: metal not detected with the chemical method.

free sample. Nevertheless, the data supplied by the biosensor were less accurate than the chemical data.

2.8 Conclusion and Future Developments

The scientific experience conducted in the laboratory UMR GEPEA over the last 10 years showed the necessity to associate biosensor technology with data analysis software. Thus, work regarding the development of a biological tool led to the design of a reliable environmental monitoring biosystem that combines a biosensor based on freeze-dried bacteria—Lumisens IV—and data analysis software—Metalsoft.

The results obtained with environmental samples showed that this technological association was able to detect and estimate more efficiently up to four metals in the mixture and environmental samples (cadmium, arsenic, mercury, and copper). Nevertheless, the estimated concentration intervals seem to be the downside of this approach. Indeed, the intervals are too wide to predict the metal concentrations in the samples as accurately as chemical measurements.

Consequently, future developments must first be devoted to the reduction of the predicted concentration intervals. For that purpose, two strategies are considered. The first consists of improving the statistical models dedicated to data analysis using other more complex mathematic algorithms (neural network, support vector machine). In parallel, the other complementary approach consists of increasing the number of bacterial sensors to obtain more accurate pollutant fingerprints.

Moreover, the actual biosystem is able to estimate only four metals in the mixture and environmental samples. Consequently, the second axis will be dedicated to the increase of the set of detected pollutants. For that purpose, the number of bioluminescent bacterial sensors must be increased. This work can take on several approaches:

- Develop other genetic constructions based on new specific inducible promoters [34–36].

- Transfer the current genetic constructions toward other bacterial strain hosts [18, 34].
- Modify the current promoter by random mutation in order to modify the sensitivity and/or specificity of the bioluminescent bacterial strains [37].

Nevertheless, the current biosystem can be used as a reliable warning tool to detect specific pollutants in environmental samples and to estimate their concentrations quickly and cost-effectively.

Acknowledgments

This research was supported by grant CER 2000-2006, Action No. 15 (Section I), research program number 18035 (Ville de la Roche sur Yon, Conseil Général de Vendée, Conseil Régional des Pays de Loire, Ministère Français chargé de la Recherche).

References

1. Kumlanghan, A., Kanatharana, P., Asawatreratanakul, P., Mattiasson, B., Thavarungkul, P. (2008). Microbial BOD sensor for monitoring treatment of wastewater from a rubber latex industry. *Enzyme Microb. Technol.*, **42**, 483–491.
2. Kress-Rogers, E. (1996). *Handbook of Biosensors and Electronic Noses*, CRC Press Book, Wauwatosa, WI, USA.
3. Charrier, T., Chapeau, C., Bendria, L., Picart, P., Daniel, P., Thouand, G. (2011). A multi-channel bioluminescent bacterial biosensor for the on-line detection of metals and toxicity. Part II: technical development and proof of concept of the biosensor. *Anal. Bioanal. Chem.*, **400**, 1061–1070.
4. D'Souza, S. F. (2001). Microbial biosensors. *Biosens. Bioelectron.*, **16**, 337–353.
5. Eltzov, E., Marks, R. S. (2010). Fiber-optic based cell sensors. *Adv. Biochem. Eng. Biotechnol.*, **117**, 131–154.
6. Eltzov, E., Marks, R. S. (2011). Whole-cell aquatic biosensors. *Anal. Bioanal. Chem.*, **400**, 895–913.
7. Lei, Y., Chen, W., Mulchandani, A. (2006). Microbial biosensors. *Anal. Chim. Acta*, **568**, 200–210.

8. Ramanathan, S., Ensor, M., Daunert, S. (1997). Bacterial biosensors for monitoring toxic metals. *Trends Biotechnol.,* **15**, 500–506.
9. Su, L., Jia, W., Hou, C., Lei, Y. (2011). Microbial biosensors: a review. *Biosens. Bioelectron.,* **26**, 1788–1799.
10. Gu, B. M., Gil, C. G. (2001). A multi-channel continuous toxicity monitoring system using recombinant bioluminescent bacteria for classification of toxicity. *Biosens. Bioelectron.,* **16**, 661–666.
11. Horry, H., Durand, M. J., Picart, P., Bendriaa, L., Daniel, P., Thouand, G. (2004). Development of a biosensor for the detection of tributyltin. *Environ. Toxicol.,* **19**, 342–345.
12. Lee, J. H., Gu, M. B. (2005). An integrated mini biosensor system for continuous water toxicity monitoring. *Biosens. Bioelectron.,* **20**, 1744–1749.
13. Horry, H., Maul, A., Thouand, G. (2007). Optimization of a bacterial bioluminescent biosensor through experimental design. *Sens. Actuators B,* **127**, 649–657.
14. Elad, T., Almog, R., Yagur-Kroll, S., Levkov, K., Melamed, S., Shacham-Diamand, Y., Belkin, S., (2011). Online monitoring of water toxicity by use of bioluminescent reporter bacterial biochips. *Environ. Sci. Technol.,* **45**, 8536–8544.
15. Pernetti, M., Poncelet, D., Thouand, G. (2009). Properties and choice of material used for microbial biosensor. In *Biosensors: Properties, Materials and Applications*, Rafael Comeaux and Pablo Novotny, New York. USA, pp. 87–126.
16. Ball, P. (1999). Applications of microporous membranes in sensor technology. *Sens. Rev.,* **19**, 46–51.
17. Roach, P. C., Ramsden, D. K., Hughes, J., Williams, P. (2003). Development of a conductimetric biosensor using immobilised Rhodococcus ruber whole cells for the detection and quantification of acrylonitrile. *Biosens. Bioelectron.,* **19**, 73–78.
18. Hakkila, K., Green, T., Leskinen, P., Ivask, A., Marks, R., Virta, M. (2004). Detection of bioavailable heavy metals in EILATox-Oregon samples using whole-cell luminescent bacterial sensors in suspension or immobilized onto fibre-optic tips. *J. Appl. Toxicol.,* **24**, 333–342.
19. Heitzer, A., Malachowsky, K., Thonnard, J. E., Bienkowski, P. R., White, D. C., Sayler, G. S. (1994). Optical biosensor for environmental on-line monitoring of naphthalene and salicylate bioavailability with an immobilized bioluminescent catabolic reporter bacterium. *Appl. Environ. Microbiol.,* **60**, 1487–1494.

20. Ivask, A., Green, T., Polyak, B., Mor, A., Kahru, A., Virta, M., Marks, R. (2007). Fibre-optic bacterial biosensors and their application for the analysis of bioavailable Hg and As in soils and sediments from Aznalcollar mining area in Spain. *Biosens. Bioelectron.,* **22**, 1396–1402.
21. Leenen, E. J. T. M., Dos Santos, V. A. P., Grolle, K. C. F., Tramper, J., Wijffels, R. (1996). Characteristics of and selection criteria for support materials for cell immobilization in wastewater treatment. *Water Res.,* **30**, 2985–2996.
22. Leth, S., Maltoni, S., Simkus, R., Mattiasson, B., Corbisier, P., Klimant, I., Wolfbeis, O. S., Osöregi, E. (2002). Engineered bacteria based biosensors for monitoring bioavailable heavy metals. *Electroanalysis,* **14**, 35–42.
23. Simpson, M. L., Sayler, G. S., Applegate, B. M., Ripp, S., Nivens, D. E., Paulus, M. J., Jellison, Jr., G. E. (1998). Bioluminescent-bioreporter integrated circuits form novel whole-cell biosensors. *Trends Biotechnol.,* **16**, 332–338.
24. Affi, M., Solliec, C., Legentillomme, P., Comiti, J., Legrand, J., Thouand, G. (2009). Numerical design of a card and related physicochemical phenomena occurring inside agarose-immobilized bacteria: a valuable tool for increasing our knowledge of biosensors. *Sens. Actuators B,* **138**, 310–317.
25. Mulchandani, A., Rogers, K. R. (1998). *Enzyme and Microbial Biosensors: Techniques and Protocols*, Humana Press, Totowa, NJ, USA.
26. Charrier, T., Durand, M. J., Jouanneau, S., Dion, M., Pernetti, M., Poncelet, D., Thouand, G. (2011). A multi-channel bioluminescent bacterial biosensor for the on-line detection of metals and toxicity. Part I: design and optimization of bioluminescent bacterial strains. *Anal. Bioanal. Chem.,* **400**, 1051–1060.
27. Horry, H., Charrier, T., Durand, M. J., Vrignaud, B., Picart, P., Daniel, P., Thouand, G. (2007). Technological conception of an optical biosensor with a disposable card for use with bioluminescent bacteria. *Sens. Actuators B,* **122**, 527–534.
28. Thouand, G., Horry, H., Durand, M. J., Picart, P., Bendriaa, L., Daniel, P., DuBow, M. S. (2003). Development of a biosensor for on-line detection of tributyltin with a recombinant bioluminescent *Escherichia coli* strain. *Appl. Microbiol. Biotechnol.,* **62**, 218–225.
29. Pernetti, M., Poncelet, D., Thouand, G., Annesini, M. C., Merli, C. (2003). Characterization of agarose as immobilization matrix model for a microbial biosensor. *Hemijskaindustrija,* **57**, 600–604.

30. Charrier, T., Durand, M. J., Affi, M., Jouanneau, S., Gezekel, H., Thouand, G. (2010). Bacterial bioluminescent biosensor characterisation for on-line monitoring of heavy metals pollutions in waste water treatment plant effluents. In *Biosensors,* Serra, P. A. (ed.), InTech, Croatia, pp. 179–206.
31. Jouanneau, S., Durand, M. J., Courcoux, P., Blusseau, T., Thouand, G. (2011). Improvement of the identification of four heavy metals in environmental samples by using predictive decision tree models coupled with a set of five bioluminescent bacteria. *Environ. Sci. Technol.,* **45**, 2925–2931.
32. Wagner, L. W., Van Dyk, T. K. (1998). Cryopreservation and reawakening. *Methods Mol. Biol.,* **102**, 123–127.
33. Jouanneau, S., Durand, M. J., Thouand, G. (2012). Online detection of metals in environmental samples: comparing two concepts of bioluminescent bacterial biosensors. *Environ. Sci. Technol.,* **46**, 11979–11987.
34. Ivask, A., Rolova, T., Kahru, A. (2009). A suite of recombinant luminescent bacterial strains for the quantification of bioavailable heavy metals and toxicity testing. *BMC Biotechnol.,* **9**, 41.
35. Monchy, S., Benotmane, M. A., Janssen, P., Vallaeys, T., Taghavi, S., van der Lelie, D., Mergeay, M. (2007). Plasmids pMOL28 and pMOL30 of *Cupriavidus metallidurans* are specialized in the maximal viable response to heavy metals. *J. Bacteriol.,* **189**, 7417–7425.
36. Verma, N., Singh, M. (2005). Biosensors for heavy metals. *BioMetals,* **18**, 121–129.
37. van der Meer, J. R., Belkin, S. (2010). Where microbiology meets microengineering: design and applications of reporter bacteria. *Nat. Rev. Microbiol.,* **8**, 511–522.

Chapter 3

Fiber Optic Biosensors for Environmental Monitoring

Evgeni Eltzov[a,b] **and Robert S. Marks**[a,b,c,d]

[a] *Department of Biotechnology Engineering, Ben-Gurion University of the Negev, Beer-Sheva 84105, Israel*
[b] *School of Material Science and Engineering, Nanyang Technology University, Nanyang Avenue, 639798, Singapore*
[c] *The National Institute for Biotechnology in the Negev, Ben-Gurion University of the Negev, Beer-Sheva 84105, Israel*
[d] *The Ilse Katz, Center for Meso and Nanoscale Science and Technology, Ben-Gurion University of the Negev, Beer-Sheva 84105, Israel*
rsmarks@bgu.ac.il

During the most recent decades, biosensors, more and more, have been entering into our lives. Despite many advantages (e.g., efficient and precise determination of all the toxicants in a sample) of analytical methodologies, whole-cell fiber optic biosensors have expanded the creation of cheaper, simpler, field operative devices. In addition to the many advantages (e.g., low price, optical measurement sensitivity, distance sensibility capabilities and absence of electrical interference) of the fiber optic transducers, the main advantage of whole-cell biosensors is the capability not only to detect the presence of bioactive compounds but also to determinate their toxic effects. In some cases it may be more important information than

Bioluminescent Microbial Biosensors: Design, Construction, and Implementation
Edited by Gérald Thouand and Robert S. Marks

ISBN 978-981-4613-65-1 (Hardcover), 978-981-4613-66-8 (eBook)
www.panstanford.com

just determination of the presence of toxicants in the environment. This chapter will review the recent advantages in whole-cell fiber optic biosensors used for monitoring air, water, and earth toxicity.

3.1 Introduction

The speed and scale of the water, earth, and air environmental total pollution rates have only deteriorated in the last decades. To induce pollution reduction, many countries around the world have in recent years implemented different environmental programs [1], not only developing countries (e.g., China where 70% of seven major rivers are severely polluted and 80% fail to meet standards for fishing [2]), to deal with severe environmental issues, such as water pollution. For example in a recent national report on water quality in the United States, 45% of assessed stream miles, 47% of assessed lake acres, and 32% of assessed bay and estuarine square miles were classified as polluted [3].

The two most popular conventional methods, gas chromatography (GC) or enzyme-linked immunosorbent assay (ELISA) [4], may provide very accurate and sensitive determination of the exact composition of testing samples [5]. But in addition to many requirements (e.g., skilled personnel, time issues, and expensive equipment) their main disadvantage is the incapability to provide a biological effect of the target analyte, while in some cases various compounds, with different structures, have a similar biological effect [6]. So in these cases this information may be much more important and relevant than the determination of the presence of this chemical substance in the tested sample. For these reasons, biosensors are good choices, which may provide cheap, fast, portable, and reliable applications for toxicity monitoring. They may measure biological (e.g., geno-/cyto-/immunotoxicity, biotoxins, endocrine) effects in different environments, such as water [7, 8], air [9], wastewater [10, 11], and soil [12, 13].

A biosensor is generally described as a self-contained, integrated receptor–transducer device capable of providing selective quantitative or semiquantitative analytical information and that uses

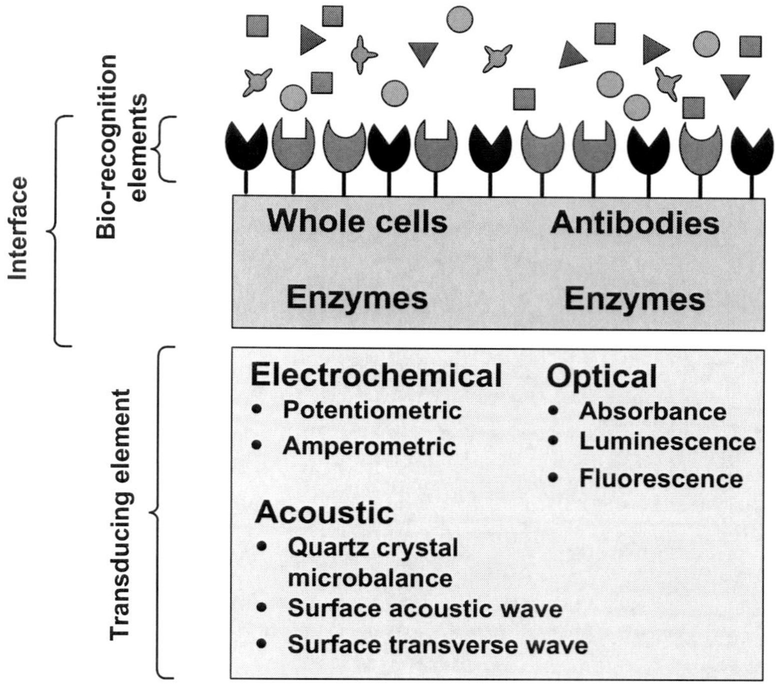

Figure 3.1 Schematic presentation of biosensors.

a biological recognition element (bioreceptor) and a transducer in intimate contact [14]. A typical biosensor is built from three different parts: bioreporter, interface, and transducer (Fig. 3.1). The bioreporter component is chosen according to the target analyte and it could be an enzyme, whole cells (e.g., virus, bacteria, yeasts, animal cells), receptors, antibodies, or biomimetic molecules. The second part is an interface where a bioreporter molecule is immobilized on the given transducer. Four different conditions should be considered—(1) maintenance of biological activity after immobilization, (2) proximity of the biological layer to the transducer, (3) stability and durability of the biological layer, and (4) sensing specificity of the biological component to a specific analyte [15]—and, for some uses, the possible future reuse of the biomaterial [16]. The principle methods of immobilization are adsorption, cross-linking, covalent binding entrapment, sol-gel entrapment and Langmuir–Blodgett (LMB) deposition self-assembled biomembranes [15, 17]. The last part of a biosensor consists of transducers that convert the recognition event to a measurable signal. The major groups of transducers are electrochemical, acoustic, piezoelectric, optical (e.g., fiber optic), mass sensitive, and thermal. More information about different biosensor and bioassay applications can be found in this handbook [18]. The ideal biosensor has to respond to low concentrations of analytes and also must be able to discriminate among species according to the biorecognition molecules that become immobilized on the interface surface of the transducing element [19]. Many researchers have engaged in detailed efforts to improve the selectivity and sensitivity of the sensing layers, together with miniaturization of the whole biosensor (e.g., probes, a related smart-signal processing system, and transduction elements) [17, 19]. The combination of optical fibers with optical sensing mechanisms and bioreporters has many advantages that make it a promising alternative to many biosensors, as well as conventional applications. Remote real-time operation, multiplexing ability, miniaturization, and immunity to electromagnetic interferences are some of the most appealing characteristics that motivate a growing scientific community [20].

Fiber optic applications frequently use bacteria or yeasts as bioreporter elements. These whole-cell organisms may be genetically tailored to produce measurable signals (e.g., light) and to respond in a specific manner to particular classes of compounds (e.g., heavy metals [21], endocrine destructive compounds (EDCs) [22, 23], antibiotics [24], and aromatic compounds [25, 26]). These whole-cell fiber optic biosensors are usually cheap and sensitive and may be easily maintained, while luminescence may be measured using a photodetector and does not suffer from a variable background signal from water or air.

In summary, fiber optic whole-cell-based biosensors allow fast and specific detection of a very wide range of toxicants and may be used as monitoring tools in different environmental and health care applications.

3.2 Fiber Optic Whole-Cell-Based Biosensors

3.2.1 *Construction of a Whole-Cell Fiber Optic Biosensor*

The ideal biosensor has to retain two main properties, that is, sensitivity (response to low concentration of analytes) and specificity (ability to discriminate among species according to the immobilized biorecognition elements). Fiber-optic whole-cell biosensors are composed of an optical fiber interfaced on one end to differently immobilized reporter organisms and on the other end to an optical photometer. Optical fibers commonly range from 80 micrometers up to several millimeters in diameter and can perform many jobs (e.g., light collection and delivery, image transmission) [27]. A typical fiber optic is built from two parts, core and cladding, while light propagates through the fiber optic due to differences in indices of refraction of these components. When the incident light is internally reflected in total, its intensity does not abruptly decay to zero at the interface. A small portion of light penetrates the reflecting medium by a fraction of the wavelength, far enough for recognition of the different refractive indexes [28]. This electromagnetic field, called the evanescent wave, has an intensity that decays exponentially with distance, starting at the interface and extending into the medium of the lower refractive index [28]. Different types of optic techniques (e.g., absorption, (bio-/chemo-/fluoroluminescence, Raman, surface plasmon resonance [SPR]) may be used in combination with fiber optic light guides. While used bioreporters (e.g., bacteria) may be placed close to the optical fiber or directly immobilized on its surface, the incorporation of an optical fiber into a biosensor results in several advantages, for example, geometric convenience and ?exibility, low cost of production, biocompatibility and no hazards, capability of remote signal transmission (i.e., remote sensing), and high-efficiency coupling in the blue-green region, which is ideal for bioluminescence ([29, 30] and can transmit multiple optical signals simultaneously, thereby offering multiplexing capabilities for sensing [31]. But in addition to the many advantages, fiber optics have two main drawbacks, incapability to simultaneously monitor two (or more) different analytes with the same fiber optic and in many cases an inability to detect the target analyte

	Advantages	Disadvantages
Transducer		
Fiber optic	• No electrical interferences • No references electrode • Not need direct contact with bioreporters • Possibilities for miniaturization • Stable • Simultaneously monitoring multi-analyte samples • Remote/ real time monitoring	• Need light produce bioreporters other light sources • Light interferences (need dark environment) • Limited dynamic range • Irreversibility • Limited stability
Immobilization method		
Sol-gel	• Chemically inert • High mechanical strength • Optical transparency • Enhanced thermostability • Biocompatibility • Resistance to microbial attack	• Only 5–20% by weight of cell mass can be incorporated • Damage bacteria • Leaching issues
Photopolymer	• May be used as structural reinforcement other matrices (e.g. alginate)	• UV damage bacteria • Some methods utilize harsh chemical initiators to facilitate the polymerization • Small pore size does not allow the capture of many cells, lowering signal response
Alginate	• Optical transparency • Provide gentle environment • Easy to create • Possibilities for modification (e.g. increasing stability or durability) • High porosity	• Instability in calcium-poor solutions and deterioration in the presence of phosphate and other calcium chelators • Low deformation resistance • Biodegradability

Figure 3.2 Advantages and disadvantages of fiber optic transducers and matrix immobilization methods.

without additional biomarkers. The first problem may be solved by the creation of a fiber optics bundle, which on each fiber can immobilize various bioreporters sensitive to different toxicants [32], while the second problem may be solved in two different ways, by creating SPR-based fiber optic biosensors or by using bioluminescent bacteria that do not need any additional additives and may produce light in the presence of pollutant [16]. Figure 3.2 summarizes some advantages and disadvantages of fiber optic biosensors.

Another critical step in creating a successful biosensor is determining the right immobilization process. In principle there are three main approaches for immobilizing bacteria on the fiber optic transducers, that is, nonspecific adsorption, chemical modification of surfaces, and direct entrapment in the gels. The immobilization step should be chosen very carefully because in some cases exposure of immobilized bacteria to the different chemicals during immobilization may damage and kill bacteria or induce and create false responses [31]. Immobilization entrapment matrices (e.g., hydrogels, sol-gels, photopolymers) suggest many advantages, for example, flexibility in creation and modification, high diffusion rates, biocompatibility, and high-water-content applications [30, 33–35]. Figure 3.2 shows the advantages and disadvantages of each proposed immobilization matrix. For example, a sol-gel is relatively inexpensive to synthesize and has several desirable properties, including chemical inertness, high mechanical strength, optical transparency, enhanced thermostability, biocompatibility, resistance to microbial attack, and the capability to immobilize bacteria at room temperature [36]. But the major disadvantage of sol-gel ceramics for cell entrapment is that only 5%–20% by weight of cell mass can be incorporated. Higher concentrations of cells reduce coating stability when hydrated [37]. Therefore their reactivity per square meter is very low and their ability to retain entrapped cell viability after storage at 4°C is poor—$<5\%$ cell viability after 9 days [38]. It has been shown that the presence of glycerol during immobilization increases overall cell viability [39]. An additional class of an immobilization matrix is a photopolymer, which uses a visible or ultraviolet light to cross-link the monomers. A photon from the light source breaks the photoinitiator into groups of highly energized radicals. The radicals then react with the resident monomer in solution and initiate the usually unstable thermoset polymerization [40].

Alginate matrices are the most used immobilization methodology for fixation of bacteria on fiber optics. In the presence of divalent ions such as calcium, alginates spontaneously form gels, while physical properties of alginates depend on the sequence of 1,4-linked β-D-mannuronic (M) and α-L-guluronic (G) acid residues as well as on the average molecular weights and molecular weight distribution of the polymer [41, 42]. In addition to the higher porosity, convenient work, and provision of a gentle environment for bacteria, the main advantage of the alginate matrices is that they can be chemically modified to improve their characteristics. Not only can they be coupled with different molecules (e.g., short peptides

[43], galactose moieties [44], poly(ethylene glycol) diamines [45], avidine–biotine [35]), but they may be modified also with other matrices (e.g., photopolymers [46] and sol-gels [47]) for better stability and functionality.

There are many other approaches for the immobilization of recombinant bacteria on solid substrates. Bacteria may be immobilized in agarose [48, 49] or with antibody-modified surfaces [50]. Despite the fact that these proposed methods were not used for optical applications, they may be easily modified for fiber optics as well.

3.2.2 *Bacterial Fiber Optic Applications*

Bacterial fiber-optic-based biosensors have been used in many different applications for toxicity measurement in air, water, and soil. This part will introduce recent advances in this field, while Table 3.1 will summarize some of the applications.

3.2.2.1 Monitoring toxic chemicals in soil

To monitor the soil toxicity with bacteria that need aquatic environments, premeasuring preparations (e.g., mixing soil with

Table 3.1 Examples of fiber optic bacterial biosensors

Target analyte	Bioreporter	Sensitivity	Ref.
Polycylic aromatic hydrocarbons (phenanthrene)	GC2 (*lac::luxCDABE*)	2 ppm	[51]
Isopropyl β-d-1-thiogalactopyranoside	*E. coli*	1.5 nM	[52]
17β-estradiol and testosterone	*S. cerevisiae*	0.5 nM	[52]
Cadmium	*Bacillus badius*	0.1 μg/L	[53]
Mercury	*E. coli*	100 nM	[32]
BTEXs (Benzene, Toluene, Ethylbenzene, and Xylene)	GC2 (*lac::luxCDABE*)	15, 16.7 72.7 and 64.9 ppm	[54]
Naphthalene	*Pseudomonas putida*	50 nM	[26]
Cooper	*Alcaligenes eutrophus*	1 μM	[55]
Tributyltin	*E. coli*	0,02 μM	[56]
Methyl parathion	*Flavobacterium*	0,3 μM	[22]

water) are required; there two main preparation and measuring methodologies, water–soil-mixed solution or supernatant of the soil–/sediment–water suspension.

A biosensor for detecting the toxicity of phenanthrene (polycylic aromatic hydrocarbon [PAH])-contaminated soil has been successfully constructed using an agar-immobilized recombinant bioluminescent bacterium, GC2 (*lac::luxCDABE*), which constitutively produces bioluminescence [51]. The occurrence of PAHs in soils and sediments is of increasing environmental concern because some PAHs exhibit mutagenic and carcinogenic effects [57]. The monitoring of phenanthrene toxicity was achieved through the measurement of the decrease in bioluminescence when a sample extracted with the biosurfactant (due to low solubility and high sorption to soil) was injected into the minibioreactor. This biosensor system using a biosurfactant may be applied as an in situ biosensor to detect the toxicity of hydrophobic contaminants in soils and for performance evaluation of PAH degradation in soils [51].

A fiber optic biosensor also may be successfully implemented in in situ bioremediation processes in order to monitor the effect of physicochemical conditions of a subsurface environment on microbial activity. For this reason a biosensor based on a *lux* bioreporter organism was created to noninvasively and in situ monitor real-time microbial activity in porous media [58]. This system not only allows the determination of the effect of physicochemical conditions in the subsurface influence microbial activity but also allows effective capture of the dynamics of in situ bacterial gene expression during naphthalene catabolism in saturated porous media.

3.2.2.2 Determining toxic chemicals in air

It is commonly accepted now that poor urban air quality can be hazardous and poses risks to people in outdoor as well as in indoor areas [59], inducing many ill effects on health, most related to nonspecific pollutant exposure. This means that the damage effect was observed after exposure to low-level air pollutants, generally not identified by a specific cause or compound [60]. So the fact that bacteria not only may be able to sense the presence of toxic compounds in the air but also may show their toxic (biological) effects makes biosensors with whole cells as bioreporters favorable to many other applications.

Humans can be exposed easily to volatile organic compounds (VOCs), for example, benzene, toluene, ethylbenzene, and xylene (BTEX), widely utilized as solvents, and they are suspected to be carcinogenic or endocrine disruptors [54]. A fiber-optic-based whole-cell biosensor was developed for the detection of toxic gaseous chemicals (e.g., BTEXs) [54]. A genetically engineered bioluminescent bacterium (*lac::luxCDABE*) was immobilized in a solid agar medium, while a decrease in the light strength due to reduction in intensity when the cells experience toxic or lethal conditions was measured with a fiber optic probe connected between the biosensor kit and a luminometer.

In another application a bioluminescent *Pseudomonas putida* naphthalene biosensor strain carrying the NAH7 plasmid (chromosomal fusion between the *sal* promoter and the *luxAB* genes) was used for measuring organic pollutants directly into the gas phase [26]. Gas-phase measurements with filter-immobilized biosensor cells in closed flasks, with a naphthalene-contaminated aqueous phase, showed that the biosensor cells can measure naphthalene effectively. It was found that cells more sensitive to naphthalene in the gas phase than in the aqueous suspension (50 nM instead of 0.5 μM) were probably due to faster transport of naphthalene to the cells in the gas phase [26].

3.2.2.3 Determining chemicals in water

A portable device based on genetically engineered bioluminescent cells (*Escherichia coli* sensitive to lactose analogue isopropyl β-d-1-thiogalactopyranoside and *Saccharomyces cerevisiae* sensitive to androgens and estrogens) was developed [52]. Cells were immobilized on a 4×3 multiwell cartridge using a new biocompatible matrix that preserved their vitality. Using a fiber optic taper, the cartridge was placed in direct contact with a cooled charge-coupled device (CCD) sensor to image and quantify the bioluminescence signals. Good stability (up to 1 month) and sensitivity (the analytes could be detected at nanomolar levels with good precision and accuracy) make this portable device a good tool for onsite toxicity monitoring. In another biosensor, methyl parathion pesticide was monitored with *Flavobacterium sp.* that was immobilized by trapping in a glass fiber filter [22]. The immobilized microbial biocomponent was disposable and cost effective and showed high reproducibility and uniformity. The detection of methyl parathion by the use of a disposable microbial biocomponent with an optical biosensor was a simple, single step and direct measurement of very low quantity (75 μL) of the sample.

Analytical methods do not always allow the correlation of the actual toxicity of the tested analyte to the total amount of heavy metals in the water. So determination of the biologically available metal fraction (bioavailability) in environmental samples is proving a crucial issue in environmental monitoring [31]. A fiber optic biosensor using sol-gel-immobilized *Bacillus badius* whole cells with phenol red as an indicator was developed and used for determining the presence of cadmium in milk [53]. As well, a heavy metal monitoring application used optical imaging fiber containing a high-density array of microwells with a single bacterium immobilized in each well for sensitive (up to 100 nM Hg^{2+}) and fast (1 h incubation time) monitoring of mercury in water [32]. Other heavy-metal-sensitive bacteria (strain of microorganisms from *Alcaligenes eutrophus*) were genetically engineered by inserting a *luxCDABE* operon from *Vibrio fischeri* under the control of a copper-induced promoter [55]. The lowest detection limit (1 μM) was achieved with microorganisms cultivated from glycerol stock solutions in the growth media and immobilized in a calcium alginate.

A novel biosensor for biocides' (e.g., tributyltin [TBT]) environmental monitoring using a bioluminescent recombinant *E. coli::luxAB* strain was developed [56]. The proposed biosensors may be divided into two parts: (1) A minibioreactor allows for both a stable and a reproducible environment for the bacterium and an in situ contact with the toxic compound, and (2) a fiber optic transducer and the analytical light device allow the real-time monitoring and analysis of the light response of bacterial cells [56]. The effect of the growth medium on bacterial sensitivity was evaluated, while a synthetic glucose medium gave a better TBT detection limit than a LB medium (respectively 0.02 μM and 1.5 μM

TBT). Another application device was built from three parts: (1) a central unit, (2) a disposable card where bioluminescent bacteria were immobilized and inserted into unit 1, and (3) an acquisition unit to control the device was used for real-time TBT monitoring in water [61]. This biosensor was able to detect 2 μM of TBT online in 400 min, and after a contact time of 1 h with the pollutant, the accumulation effect of the matrix allowed us to detect 1 nM of this toxic biocide.

3.3 Creation of a Portable Biosensor for Toxicity Monitoring

3.3.1 *Determining Chemicals in Water*

In this part the creation of fiber optic bacterial biosensors for the detection of different toxicants in static as well as in flow-through water and air will be described. Figure 3.3 shows details of these sensors, while in all described applications was included a photon-counting system to measure a low-light response produced by recognition elements (Fig. 3.3A) [30]. This photodetector (Hamamatsu HC135-01 photomultiplier tube [PMT] sensor module) has an optimum blue light region and includes a 21 mm diameter active area for gathering light radiation without any optical focusing elements. To prevent light disturbances and damaging of the photon-counting unit, the instrument setup was placed in a light-tight box and a manual shutter was placed in front of the detector (Fig. 3.3A). To receive and treat data, a specific driver was developed using LabView (Version 3.1, National Instruments Corporation), which allows the monitoring of the luminescent signal and data handling in real time. One end of the fiber optic was placed in front of the PMT, while on other side bioluminescent bacteria sensitive to the different toxicants or stresses were immobilized (Fig. 3.3B). Bacterial cells are ideal bioreporters for environmental monitoring because of their low cost, fast growth, rich genetic modifications, easy handling, and sensitivity to a wide variety of environmental stimuli [62]. As referred to above, the main advantage of bacteria is the capability for genetic modifications. This means that bacteria

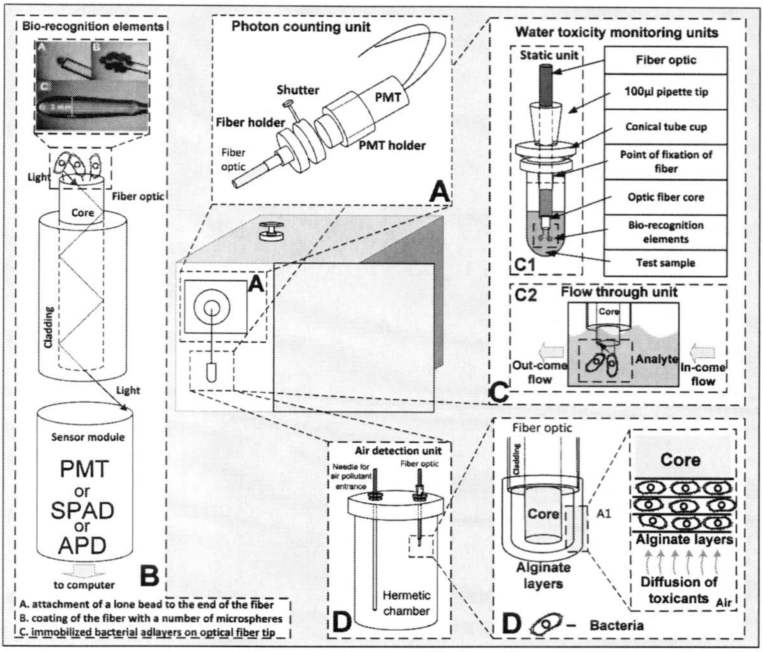

Figure 3.3 Fiber optic bacterial biosensor. (A) Measuring unit; (B) schematic presentation of the fiber optic with immobilized bacteria; (C) water measurement mode, C1: static; C2; flow-through; and (D) air measurement mode.

may be genetically modified to sense selectively target analytes in different environments. Table 3.2 shows different bacterial strains sensitive to various toxicants.

The alginate immobilization technology was used for fixation of bacteria on the fiber optic. Two different approaches were used. In the first, bacteria were immobilized in the calcium alginate layers. The layers were created by dipping the fiber optic into a mixed solution of a reporter bacteria and filter-sterilized 2% (w/v) sodium alginate (1:1 ratio) and then into sterile 0.5 M calcium chloride solution. Repeating these steps thickened the adlayer, thus increasing the number of bacterial cell sensors attached to the optical fiber transducer. A DPD1718 strain sensitive to genotoxic stresses was immobilized on the fiber optics and exposed to the toxicants [30]. It was demonstrated that immobilized

Table 3.2 Examples of different bioluminescent bacterial strains

Bacterial strain	Promoter	Target analyte	Ref.
DPD2794	*recA*	Genotoxicity	[63]
DPD1718	*recA*	Genotoxicity	[64]
TV1061	*grpE*	Cytotoxicity	[65]
K802NR	*lasI*	Quorum sensing	[24]
MC1061	*pmerR*	Hg	[66]
MC1061	*Pars*	As	[67]
BR151,RN4220	*cadA*	Cd, Ld	[68]
MC1061	*copA*	Cu	[21]
DPD2511	*katG*	Oxidative stresses	[50]
DPD2544	*fabA*	Fatty acid dependence	[50]
R306	*ORF6-pefI*	Quorum sensing	[77]

bioluminescent bacteria can produce light in response to an external inducing agent in a dose-dependent manner, while the light response was diffusion rate dependent. Six layers were determined as an optimal number. Future optimizations show that many other parameters (e.g., bacterial cell density, numerical aperture of the fiber core, working temperature) could be optimized for increasing biosensor sensitivity and for reducing threshold [34]. Using this immobilization technology and immobilized recombinant *E. coli* cells carrying genes responsive to the presence of bioavailable heavy metal ions (e.g., Hg^{2+}, Cu^{2+}, Cd^{2+}, and Pb^{2+}) with fusion to a firefly luciferase (*luc*) reporter gene (Table 3.2) 17 synthetic and 3 environmentally blind samples were tested in the presence of heavy metals [21] (Fig. 3.3C1). This study not only showed the capability of this bacterial sensor to quantify bioavailable metals from unknown samples but also demonstrated a simplicity of the measurements (samples do not need any pretreatment before analysis) and field usage possibility. The second immobilization strategy was based on alginate–biotin microspheres with encapsulated bioluminescent reporter cells. A biosensor was prepared by conjugating these biotinylated alginate microspheres to the surface of a streptavidin-coated optical fiber, while the performance of the biosensor was demonstrated in the determination of the antibiotic mitomycin C as a model toxin [35] (Fig. 3.3B). The bond between biotin and avidin is established very rapidly and, once formed, is essentially undisturbed by extreme pH values or exposure to organic solvents or other denaturating agents [70]. The main advantage of this new immobilization methodology is the combination of the advantages of the alginate's gelling properties to entrap cells, providing them with a gentle hydrated and highly porous environment, and the high affinity interaction of the avidin–/streptavidin–biotin system [35].

All biosensors described above allow the detection of water toxicity when static. This means that water samples have to be taken to the testing chamber from the main source and only then can it be measured. The main disadvantage of these applications is the probability that in the moment of water collected, the toxicant will not be inside the collected sample (e.g., in flowing water, such as rivers or tap water, where exposure to pollutants may be time dependent). So there is a huge demand for the development of a simple field monitor application for real-time measurements of water toxicity. Such a new system for online monitoring of toxic pollutants in water has been proposed [8]. To monitor toxicity in real time, a field-operable fiber optic photodetector device was designed in a flow-through manner (Fig. 3.3C2). Fiber-optic-alginate-immobilized bacteria were placed in flowing water and exposed for a half hour to toxic compounds in the water stream (mytomicin C or *p*-chlorophenol). The result showed a capability of this biosensor to continuously monitor the presence of toxic compounds in a dose-dependent manner, where increasing the pollutant concentration increased the light response of the bioreporter bioluminescent bacteria. Twenty-four-hour measurements in tap water demonstrated the ability of the device to run for such a time period. Sensitivity of this application was roughly comparable to that of the static fiber optic biosensors [30, 34, 71]. This flow-through fiber-optic-based monitoring system has proven to be a useful next step in the development of a simple online real-time sensor for relevant human toxicants in flowing water [8].

3.3.2 *Determining Chemicals in Air*

Air pollution is a serious problem in many developed countries. But little attention is paid to air pollution in homes and office buildings, and for many years, such environments were assumed free of health-damaging pollutants. In the last two decades, the tendency to accept this view has changed. Now it is well known that poor urban air quality may be hazardous and may pose a risk to people in indoor areas [59], inducing many ill effects on health, mainly related to the exposure to low-level air pollutants, generally not identified as a specific cause or compound [60]. In light of the obvious need for real-time air pollution monitoring, a fiber optic whole-cell biosensor for real-time monitoring of air toxicity was proposed [9]. At the beginning, two bioluminescent bacterial reporter strains (genotoxicity and cytotoxicity sensitive) were immobilized in different matrices and exposed to the air pollutant, while many parameters (e.g., type of matrix, volume and type of bacteria) were tested. The TV1061 strain immobilized in an alginate showed the best sensitivity to the tested compound and direct correlation between the volume of the immobilization bed and corresponding bacterial response and viability was found. In consequence, a novel online fiber-optic-based biosensor was created (Fig. 3.3D). A TV1061 strain was immobilized with an alginate on the proximal end of a fiber optic and exposed to different air pollutants (e.g., chloroform, toluene, and compounds originating from paints). Three layers of alginate were determined to be optimum for air monitoring. The sensor was able to show pollutants having differences in strength and shape in the bioluminescent response. This sensor could be useful mostly as an early warning system in office and home areas, as well as in outdoor environments.

3.3.3 *Determining Chemicals in Soil*

Because a number of processes influence the bioavailability of contaminants in soils and sediments, determining bioavailability in these matrices is a very complicated task, and thus far no universal test has been developed [72]. Different bioluminescent whole-cell bioreporters have been used for monitoring the presence of various bioavailable heavy metals (e.g., arsenate [67, 73], mercury [74], cadmium [75], and copper [76]) in solid samples, such as soils and sediments. But all these applications need lab equipment and cannot be used as field (onsite) biosensors. So to create a field portable device, two bioluminescent bacteria were immobilized on a fiber optic and used for monitoring mercury and arsenate in the soil samples [66]. The optimized biosensors (consisting of seven layers of fiber-attached bacteria pregrown until the midlogarithmic growth phase) enabled quantification of environmentally relevant concentrations of the target analytes: 2.6 μg/L of Hg(II) and 141 μg/L of As(V) or 18 μg/L of As(III). Furthermore, the importance of the direct contact of bacteria to the soil analytes was demonstrated, while alginate-immobilized bacteria showed 20-fold less Hg and 4-fold less As responses than nonimmobilized bioreporter bacteria. In the case of Hg, complexation of the Hg ions with the alginate matrix could also reduce the bioavailable fraction. This biosensor application can successfully complement the analytical chemical methods in environmental risk assessment.

3.3.4 *Improving the Immobilization Methodology*

In the sections above, fiber optic biosensors based on bioluminescent bacteria immobilized in alginate matrices and used to monitor dangerous toxicants in air, water, and soil have been show. But in addition to the many advantages (e.g., convenient environment for fixed cells, compounds inside and outside fast diffusion rates), alginates used as immobilization matrices were very sensitive to environmental conditions (e.g., chelators or physical damage) and were not able to prevent cell leakage outside the matrices. This important issue, because of the usage of genetically modified organisms (GMOs) as bioreporters in the European Union, is a strict regulation that allows using these organisms only in licensed (MLI or II) laboratories. So companies using GMOs outside special laboratories have to obtain a license for specified field use of these sensors under the condition that all tested samples will be sterilized and controlled regularly for the complete absence of living GMOs. To increase the physical strength of the matrix and to prevent GMO leakage, the fiber optics were modified with additional layers of polysulfone (Fig. 3.4). To decrease the toxicity of the modification process, optimal conditions were determined (e.g., four internal layers of alginate with bacteria and two external layers with alginate only and 2 seconds of exposure to -180°C prior to the polysulfone

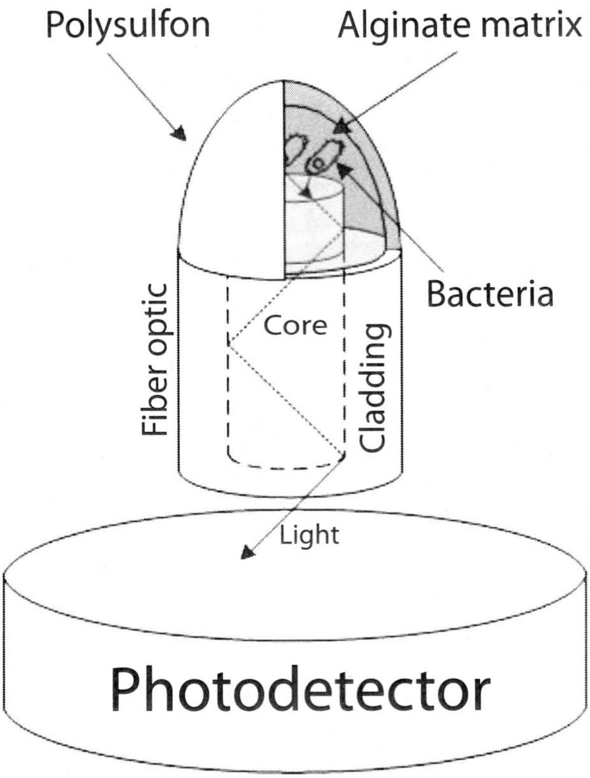

Figure 3.4 Design of a fiber optic with additional layers of polysulfone to increase the physical strength of the matrix.

addition). To determine the sensitivity of the new "modified" fibers, they were exposed to environmental toxicants and compared to fibers with alginate only. Two environmental toxicants were used, atrazine and mitomycin C. New modifications allowed stronger cell response to these toxicants and thus the highest sensitivity of 10 pg/mL for atrazine and 0.008 ppb for mitomycin C. The possible reason for this phenomenon is a synergetic effect of the fiber treatment during polysulfone placement, which made immobilized cells more susceptible to environmental toxicants. The advantage of this procedure was not only in preventing the GMO leakage outside the matrix but also in increasing sensor sensitivity by inducing cell response to lower environmental toxicants (Eltzov et al., unpublished).

3.4 Conclusions and Future Perspectives

The fast response time, cost-effectiveness, sensitivity, and simplicity of implementation make bacterial-based biosensors extremely attractive in other numerous toxicity assays. But there are still a lot of different issues that must be considered to facilitate widespread acceptance of fiber optic bacterial methods for environmental applications. For example, many bioreporter bacteria that are used in biosensors are genetically modified. For European markets (which have very strict laws forbidding exposure and release of GMOs into the environment) immobilization methods have to be significantly improved to totally prevent leakage of any cells into the environment. So future developments will be concentrating on the creation of novel immobilization methods, which in one way will allow gentle/moderate support to cells with better stability and diffusion rates and also will totally prevent leakage of the cells into the environment. Next, fiber optic sensors must facilitate the multianalyte measuring options. The main disadvantage of all the described biosensors is an inability to measure several compounds at the same time (in static or real-time monitoring modes). So in addition to future progress in genetics that will see the creation of novel bioreporters, more sensitive and specific, efforts will be made to develop technologies that allow real-time, simultaneous monitoring of several toxicants with the same measurement procedure. The future sensors may be implemented in some potential markets that include monitoring of groundwater and surface water, wastewater treatment plants, and wells continuously around waste storage facilities.

Another possible area of future improvements will be in continuous miniaturization of whole-cell fiber optic biosensors, while there are many challenges in the development of such a portable device, for example, amplification of the signal intensity to indicate target analytes and increased sensitivity and development and optimization of new, smaller components of biosensors (pumps, filters, membranes, sample handling, and conditioning). As well, with evolving nanotechnologies, new efforts will be focused on implementation of these technologies in the fiber optic field—for example, creation of silver nanoislands on the fiber optic to increase light responses of bacteria that will eventually increase sensitivity [77].

To summarize, biosensors, the technology of the future, will more and more intrude in our lives as routine monitoring devices and will better prevent exposure of people to toxic environments.

References

1. Yang, Y. S., Wang, L. (2010). A review of modelling tools for implementation of the eu water framework directive in handling diffuse water pollution. *Water Resour. Manage.*, **24**(9), 1819–1843.
2. Zhang, G. X., Deng, W. (2002). The groundwater crisis and sustainable agriculture in northern China. *Water-Eng. Manage.*, **149**(4), 13–16.
3. United States Environmental Protection Agency. (2007). *The National Water Quality Inventory: Report to Congress for the 2002 Reporting Cycle; A Profile.* United States Environmental Protection Agency, Washington, DC, USA.
4. De Meulenaer, B., Baert, K., Lanckriet, H., Van Hoed, V., Huyghebaert, A. (2002). Development of an enzyme-linked immunosorbentassay for bisphenol a using chicken immunoglobulins. *J. Agric. Food Chem.*, **50**(19), 5273–5282.
5. Belkin, S. (2003). Microbial whole-cell sensing systems of environmental pollutants. *Curr. Opin. Microbiol.*, **6**(3), 206–212.
6. Sonnenschein, C., Soto, A. M. (1998). An updated review of environmental estrogen and androgen mimics and antagonists. *J. Steroid Biochem. Mol. Biol.*, **65**(1–6), 143–150.
7. Eltzov, E., Marks, R. S. (2011). Whole-cell aquatic biosensors. *Anal. Bioanal. Chem.*, **400**(4), 895–913.
8. Eltzov, E., Marks, R. S., Voost, S., Wullings, B. A., Heringa, M.B. (2009). Flow-through real time bacterial biosensor for toxic compounds in water. *Sens. Actuators B*, **142**(1), 11–18.
9. Eltzov, E., Pavluchkov, V., Burstin, M., Marks, R. S. (2011). Creation of a fiber optic based biosensor for air toxicity monitoring. *Sens. Actuators B*, **155**(2), 859–867.

10. Raud, M., et al. (2010). Semi-specific biosensors for measuring BOD in dairy wastewater. *J. Chem. Technol. Biotechnol.*, **85**(7), 957–961.

11. Ponomareva, O., et al. (2011). Microbial biosensors for detection of biological oxygen demand (a review). *Appl. Biochem. Microbiol.*, **47**(1), 1–11.

12. Maderova, L., Watson, M., Paton, G. I. (2011). Bioavailability and toxicity of copper in soils: integrating chemical approaches with responses of microbial biosensors. *Soil Biol. Biochem.*, 43(6), 1162–1168.

13. Donner, E., et al. (2010). Biological and chemical assessments of zinc ageing in field soils. *Environ. Pollut.*, **158**(1), 339–345.

14. Thevenot, D. R., et al. (2001). Electrochemical biosensors: recommended definitions and classification. *Biosens. Bioelectron.*, **16**(1–2), 121–131.

15. Collings, A. F., Caruso, F. (1997). Biosensors: recent advances. *Rep. Prog. Phys.*, **60**(11), 1397–1445.

16. D'Souza, S. F. (2001). Microbial biosensors. *Biosens. Bioelectron.*, **16**(6), 337–353.

17. Eltzov, E., Kushmaro, A., Marks, R. S. (2009). Biosensors and related techniques for endocrine disruptors. In *Endocrine Disrupting Chemicals in Food*, Snow, I. (ed.), Woodhead, Cambridge, UK.

18. Marks, R. S., et al. (2007). *Handbook of Biosensors and Biochips*, John Wiley and Sons, Oxford.

19. Eltzov, E., Marks, R. S. (2009). Parameters to consider in the construction of fiber-optic biosensors as alternative bioanalytical tools. *IEEE Instrum. Meas. Mag.*, **12**(5), 10–16.

20. Jorge, P. A. S. (2010). Fiber optic refractometric configurations for environmental sensing. In *Proceedings of the 3rd WSEAS International Conference on Advances in Sensors, Signals and Materials*, World Scientific and Engineering Academy and Society (WSEAS), Faro, Portugal, pp. 174–178.

21. Hakkila, K., Green, T., Leskinen, P., Ivask, A., Marks, R., Virta, M. (2004). Detection of bioavailable heavy metals in EILATox-Oregon samples using whole-cell luminescent bacterial sensors in suspension or immobilized onto fibre-optic tips. *J. Appl. Toxicol.*, **24**(5), 333–342.

22. Kumar, J., Jha, S. K., D'Souza, S. F. (2006). Optical microbial biosensor for detection of methyl parathion pesticide using Flavobacteriumsp whole cells adsorbed on glass fiber filters as disposable biocomponent. *Biosens. Bioelectron.*, **21**(11), 2100–2105.

23. Fine, T., et al. (2006). Luminescent yeast cells entrapped in hydrogels for estrogenic endocrine disrupting chemical biodetection. *Biosens. Bioelectron.*, **21**(12), 2263–2269.

24. Eltzov, E., et al. (2008). Detection of sub-inhibitory antibiotic concentrations via luminescent sensing bacteria and prediction of their mode of action. *Sens. Actuators B*, **129**(2), 685–692.

25. Lee, H. J., et al. (2003). Monitoring and classification of PAH toxicity using an immobilized bioluminescent bacteria. *Biosens. Bioelectron.*, **18**(5–6), 571–577.

26. Werlen, C., Jaspers, M. C. M., van der Meer, J. R. (2004). Measurement of biologically available naphthalene in gas and aqueous phases by use of a Pseudomonas putida biosensor. *Appl. Environ. Microbiol.*, **70**(1), 43–51.

27. Flusberg, B. A., et al. (2005). Fiber-optic fluorescence imaging. *Nat. Methods*, **2**(12), 941–950.

28. Marazuela, M. D., Moreno-Bondi, M. C. (2002). Fiber-optic biosensors: an overview. *Anal. Bioanal. Chem.*, **372**(5–6), 664–682.

29. Marks, R. S., et al. (1997). Chemiluminescent optical fiber immunosensor for detecting cholera antitoxin. *Opt. Eng.*, **36**(12), 3258–3264.

30. Polyak, B., et al. (2000). Optical fiber bioluminescent whole-cell microbial biosensors to genotoxicants. *Water Sci. Technol.*, **42**(1–2), 305–311.

31. Eltzov, E., Marks, R. S. (2010). Fiber-optic based cell sensors. *Adv. Biochem. Eng. Biotechnol.*, **117**, 131–154.

32. Biran, I., et al. (2003). Optical imaging fiber-based live bacterial cell array biosensor. *Anal. Biochem.*, **315**(1), 106–113.

33. Elisseeff, J., et al. (2000). Photoencapsulation of chondrocytes in poly(ethylene oxide)-based semi-interpenetrating networks. *J. Biomed. Mater. Res.*, **51**(2), 164–171.

34. Polyak, B., et al. (2001). Bioluminescent whole cell optical fiber sensor to genotoxicants: system optimization. *Sens. Actuators B*, **74**(1–3), 18–26.

35. Polyak, B., Geresh, S., Marks, R. S. (2004). Synthesis and characterization of a biotin-alginate conjugate and its application in a biosensor construction. *Biomacromolecules*, **5**(2), 389–396.

36. Yu, D., et al. (2005). Aqueous sol-gel encapsulation of genetically engineered Moraxella spp. cells for the detection of organophosphates. *Biosens. Bioelectron.*, **20**(7), 1433–1437.

37. Flickinger, M. C., et al. (2007). Painting and printing living bacteria: engineering nanoporous biocatalytic coatings to preserve microbial viability and intensify reactivity. *Biotechnol. Prog.*, **23**(1), 2–17.

38. Bottcher, H., et al. (2004). Biocers: ceramics with incorporated microorganisms for biocatalytic, biosorptive and functional materials development. *J. Mater. Chem.*, **14**(14), 2176–2188.

39. Nassif, N., et al. (2003). A sol-gel matrix to preserve the viability of encapsulated bacteria. *J. Mater. Chem.*, **13**, 203–208.

40. Bryant, S. J., Nuttelman, C. R., Anseth, K. S. (2000). Cytocompatibility of UV and visible light photoinitiating systems on cultured NIH/3T3 fibroblasts in vitro. *J. Biomater. Sci. Polym. Ed.*, **11**(5), 439–457.

41. Tombs, M., Harding, S. E. (1998). *An Introduction to Polysaccharide Biotechnology*, Taylor and Francis, London, UK.

42. Gacesa, P. (1998). Alginates. *Carbohydr. Polym.*, **8**, 161–182.

43. Rowley, J. A., Madlambayan, G., Mooney, D. J. (1999). Alginate hydrogels as synthetic extracellular matrix materials. *Biomaterials*, **20**(1), 45–53.

44. Yang, J., et al. (2002). Galactosylated alginate as a scaffold for hepatocytes entrapment. *Biomaterials*, **23**(2), 471–479.

45. Eiselt, P., Lee, K. Y., Mooney, D. J. (1999). Rigidity of two-component hydrogels prepared from alginate polyethylene glycol diamines. *Macromolecules*, **32**, 5561–5566.

46. Lu, M. Z., et al. (2000). Cell encapsulation with alginate and alpha-phenoxycinnamylidene-acetylated poly(allylamine). *Biotechnol. Bioeng.*, **70**(5), 479–483.

47. Coradin, T., Nassif, N., Livage, J. (2003). Silica-alginate composites for microencapsulation. *Appl. Microbiol. Biotechnol.*, **61**(5–6), 429–434.

48. Bettaieb, F., et al. (2007). Immobilization of *E. coli* bacteria in three-dimensional matrices for ISFET biosensor design. *Bioelectrochemistry*, **71**(2), 118–125.

49. Alkorta, I., et al. (2006). Bioluminescent bacterial biosensors for the assessment of metal toxicity and bioavailability in soils. *Rev. Environ. Health*, **21**(2), 139–152.

50. Premkumar, J. R., et al. (2001). Antibody-based immobilization of bioluminescent bacterial sensor cells. *Talanta*, **55**(5), 1029–1038.

51. Gu, M. B., Chang, S. T. (2001). Soil biosensor for the detection of PAH toxicity using an immobilized recombinant bacterium and a biosurfactant. *Biosens. Bioelectron.*, **16**(9–12), 667–674.

52. Roda, A., et al. (2011). A portable bioluminescence engineered cell-based biosensor for on-site applications. *Biosens. Bioelectron.*, **26**(8), 3647–3653.

53. Verma, N., Kumar, S., Kaur, H. (2011). Fiber optic biosensor for the detection of Cd in milk. *J. Biosens. Bioelectron.*, **1**, 102. doi:10.4172/2155-6210.1000102.
54. Gil, G. C., Kim, Y. J., Gu, M. B. (2002). Enhancement in the sensitivity of a gas biosensor by using an advanced immobilization of a recombinant bioluminescent bacterium. *Biosens. Bioelectron.*, **17**(5), 427–432.
55. Leth, S., et al. (2002). Engineered bacteria based biosensors for monitoring bioavailable heavy metals. *Electroanalysis*, **14**(1), 35–42.
56. Thouand, G., Horry, H., Durand, M. J., Picart, P., Bendriaa, L., Daniel, P., DuBow, M. S. (2003). Development of a biosensor for on-line detection of tributyltin with a recombinant bioluminescent *Escherichia coli* strain. *Appl. Microbiol. Biotechnol.*, **62**(2–3), 218–225.
57. Gu, M. (2005). Environmental biosensors using bioluminescent bacteria. In *Environmental Chemistry*, Lichtfouse, E., Schwarzbauer, J., Robert, D. (eds.), Springer, Berlin, Heidelberg, pp. 691–698.
58. Dorn, J. G., et al. (2004). Employing a novel fiber optic detection system to monitor the dynamics of in situ lux bioreporter activity in porous media: system performance update. *Anal. Chim. Acta*, **525**(1), 63–74.
59. Brickus, L. S. R., Cardoso, J. N., Neto, F. R. D. (1998). Distributions of indoor and outdoor air pollutants in Rio de Janeiro, Brazil: implications to indoor air quality in bayside offices. *Environ. Sci. Technol.*, **32**(22), 3485–3490.
60. Mølhave, L. (2003). Organic compounds as indicators of air pollution. *Indoor Air*, **13**, 12–19.
61. Horry, H., et al. (2007). Technological conception of an optical biosensor with a disposable card for use with bioluminescent bacteria. *Sens. Actuators B*, **122**(2), 527–534.
62. Suo, Z., et al. (2008). Efficient immobilization and patterning of live bacterial cells. *Langmuir*, **24**(8), 4161–4167.
63. Vollmer, A. C., et al. (1997). Detection of DNA damage by use of Escherichia coli carrying recA'::lux, uvrA'::lux, or alkA'::lux reporter plasmids. *Appl. Environ. Microbiol.*, **63**(7), 2566–2571.
64. Davidov, Y., et al. (2000). Improved bacterial SOS promoter:: lux fusions for genotoxicity detection. *Mutat. Res.-Genet. Toxicol. Environ. Mutagen.*, **466**(1), 97–107.
65. Van Dyk, T. K., et al. (1994). Rapid and sensitive pollutant detection by induction of heat shock gene-bioluminescence gene fusions. *Appl. Environ. Microbiol.*, **60**(5), 1414–1420.

66. Ivask, A., et al. (2007). Fibre-optic bacterial biosensors and their application for the analysis of bioavailable Hg and As in soils and sediments from Aznalcollar mining area in Spain. *Biosens. Bioelectron.*, **22**(7), 1396–1402.

67. Turpeinen, R., Virta, M., Häggblom, M. M. (2003). Analysis of arsenic bioavailability in contaminated soils. *Environ. Toxicol. Chem.*, **22**(1), 1–6.

68. Tauriainen, S., et al. (1998). Luminescent bacterial sensor for cadmium and lead. *Biosens. Bioelectron.*, **13**(9), 931–938.

69. Bindhu, M., et al. (2001). SdiA of Salmonella entericals a LuxR homolog that detects mixed microbial communities. *J. Bacteriol.*, **183**(19), 5733–5742.

70. Green, N. M. (1975), Avidin. In *Advances in Protein Chemistry*, Anfinsen, J. J. T. E. C. B., Frederic, M. R. (eds.), Academic Press, pp. 85–133.

71. Pedahzur, R., et al. (2004). Water toxicity detection by a panel of stress-responsive luminescent bacteria. *J. Appl. Toxicol.*, **24**(5), 343–348.

72. Hynninen, A., Virta, M. (2010). Whole-cell bioreporters for the detection of bioavailable metals. *Adv. Biochem. Eng. Biotechnol.*, **118**, 31–63.

73. Petänen, T., et al. (2003). Assessing sediment toxicity and arsenite concentration with bacterial and traditional methods. *Environ. Pollut.*, **122**(3), 407–415.

74. Fu, Y.-J., Chen, W.-L., Huang, Q.-Y. (2008). Construction of two lux-tagged Hg2+-specific biosensors and their luminescence performance. *Appl. Microbiol. Biotechnol.*, **79**(3), 363–370.

75. Ivask, A., et al. (2004). Recombinant luminescent bacterial sensors for the measurement of bioavailability of cadmium and lead in soils polluted by metal smelters. *Chemosphere*, **55**(2), 147–156.

76. Brandt, K. K., Holm, P. E., Nybroe, O. (2006). Bioavailability and toxicity of soil particle-associated copper as determined by two bioluminescent Pseudomonas fluorescens biosensor strains. *Environ. Toxicol. Chem.*, **25**(7), 1738–1741.

77. Eltzov, E., et al. (2009). Metal-enhanced bioluminescence: an approach for monitoring biological luminescent processes. *Appl. Phys. Lett.*, **94**(8), 083901-1–083901-3.

Chapter 4

Non-Fiber-Optic Bioluminescent Biosensors

Phillip Myer,[a] Maria del Busto-Ramos,[b] Lisa Hartono,[c] and Bruce Applegate[a,b]

[a] *Department of Biological Sciences, Purdue University, West Lafayette, IN, USA*
[b] *Department of Food Science, Purdue University, West Lafayette, IN, USA*
[c] *Department of Applied and Biological Engineering, Purdue University, West Lafayette, IN, USA*
pmyer@purdue.edu

4.1 Introduction

Biological sensing elements, or biosensors, have been used to monitor gene expression in many applications and scenarios [1, 2]. Their areas of application seem only constrained by scientific creativity. Yet, in terms of practicality, many sensors require additional information, such as determination of enzyme activity, continuous sampling for activity evaluation, or even the addition of exogenous reagents. In this regard, the use of certain biosensors can be time consuming, intricate, and cumbersome. A truly viable biosensor would include elements that are rapid, noninvasive, and sensitive and can be conducted as real-time reporters of probed activity.

Bioluminescent Microbial Biosensors: Design, Construction, and Implementation
Edited by Gérald Thouand and Robert S. Marks

ISBN 978-981-4613-65-1 (Hardcover), 978-981-4613-66-8 (eBook)
www.panstanford.com

4.2 Biosensors

Biosensors comprise two components, a biological sensing/recognition element and a signal transducer. Many biological elements have been used in this regard. These elements can range from enzymes to receptors, nucleic acids, organelles, antibodies, and whole-cell reporters, for example. A competent sensing system must contain characteristics that make it specific, reliable, repeatable, able to develop real-time data, and, in some specific cases, portable [3]. In this regard, the biological signaling element can be used in its native form, or a genetically/biochemically altered version can be employed to develop a more sensitive probe capable of targeting specific analytes. In this genomic age, the latter has become increasing popular. Yet, each of these biological elements is dependent upon interfacing the specific signal to a transducer, which can monitor and quantify the biological signal. Commonly, transducers have included electrochemical, thermal, magnetic, or optical interfaces in order to convert the biological input to a digital output. The type of transducer is dependent on the signaling element utilized.

Over the past decade, bioluminescent biosensors have been used in a wide variety of assays and studies, including monitoring of cellular toxicity, bioavailability of pollutants, and DNA-damaging agents [4–7]. The bioluminescent system exploited can differ dependent upon many factors. The most commonly used whole-cell bioluminescent reporters are listed in Table 4.1. In these cases, bioluminescent reporter systems based on light emission from bacteria have been advantageous in examining biological systems due to their sensitivity, rapid detection, and noninvasiveness. For supplemental information regarding these systems, see the literature referenced in Table 4.1. Note that the reporters listed do not include any fluorescent proteins such as green fluorescent protein (GFP). This review focuses only on bioluminescent reporters, ones not requiring excitation in order to elicit a light signal.

The development of capable biosensors is dependent upon coupling the biological sensing element with an appropriate transducer. Bioluminescent biosensors commonly couple the biological input with an element able to transfer the light, augment

Table 4.1 Commonly used *lux* gene cassettes for bioluminescent-based reporting

System	Origin	Wavelength emitted (nm)
Luciferase (Lux) [8]	Bacterial: *Vibrio, Photobacterium, Photorhabdus*	490
Luciferase (Luc) [8]	Insects: Predominantly *Photinus pyralis*	562
Aequorin [8, 9]	Luminescent jellyfish: *Aequorea aequorea*	470

the signal, develop a digital response, and record the response. In many cases, this involves fiber optics, photomultiplier tubes (PMTs), and data acquisition systems. But with the increasing application of bioluminescent reporter systems across many fields, the variability in these systems has also increased to accommodate for environmental applications, specificity, size, detection limits, and even novel technologies. Due to the size and scope of applications of the biosensing element, and outside of genetically manipulating the element, the component in the system most applicable to change and enhancement is the transducer. Changes and advancements in technology regarding the interfacing of the biological input with the digital output has provided scientists with many tools to examine systems, not only with specificity, but also in real time and with automation. The following sections will discuss the different types and applications of whole-cell bioluminescent biosensor systems with regard to abiotic components. Additionally, due to common applications and knowledge of fiber optics in these systems, the focus will cover non-fiber-optic systems.

4.3 Biosensor Light Throughput

One of the most important aspects of the *luxCDABE* cassette-based bacterial bioluminescence system is its autonomous nature [10]. Because of its autonomous expression system, it is easily interfaced with signal transducers. This has proven important in many instances of applications due to this ease in configuration. Bioluminescent reporters are immobilized and light can be transferred directly to a detection system. One of the most common instruments

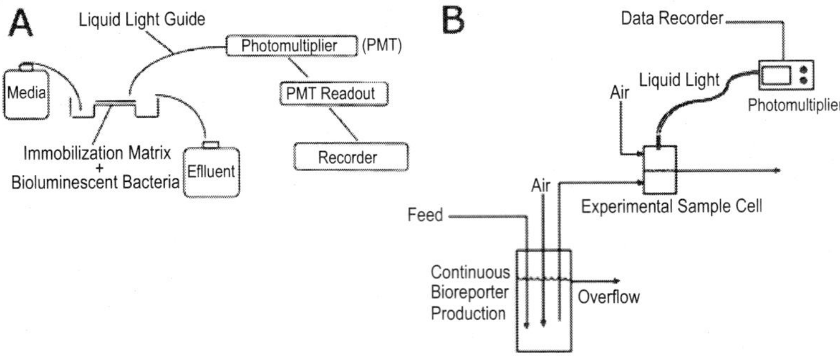

Figure 4.1 Examples of biosensors utilizing liquid light guides. (A) Schematic of an immobilized-biosensor light-monitoring system. Adapted from Elasri et al. (2000) [4]. (B) Schematic of a continuous sampling system. Adapted from Kelly et al. (1999) [12].

of light throughput in these systems is fiber optic cables. These optical fiber bundles transmit light on the basis of the principal of total internal reflection (TIR). Light is guided through the fibers with very little loss to the surroundings. They are easily applied when the bioreporter is immobilized.

Alternatively, liquid light guides have grown popular in recent years, especially with reference to more dynamic systems. In comparison to fused silica fiber bundles, liquid light guides are more efficient, less expensive, and less fragile; breakage is a nonissue [11]. As with fiber-optic-based light throughput modules, liquid light guides offer the same application potential. For example, they have been used in studies in which the bioreporter was immobilized (Fig. 4.1A) [4].

In these systems, bioreporter bacteria are immobilized (e.g., an alginate matrix), a testing stimulus or agent is applied, and the bioluminescent response is transferred from the immobilized reporter to a recorder via the liquid light guide. In addition, liquid light guides have been utilized in continuous systems, where bioreporter production is continuously fed into the experimental/control light-sampling cells (Fig. 4.1B) [12].

Additionally, Hartano et at. (2008), created a biosensor based on a bioluminescent *Escherichia coli* O157:H7 (*E. coli* O157:H7-*lux*) [13]. *E. coli* O157:H7-*lux* was constructed using a plasmid-based

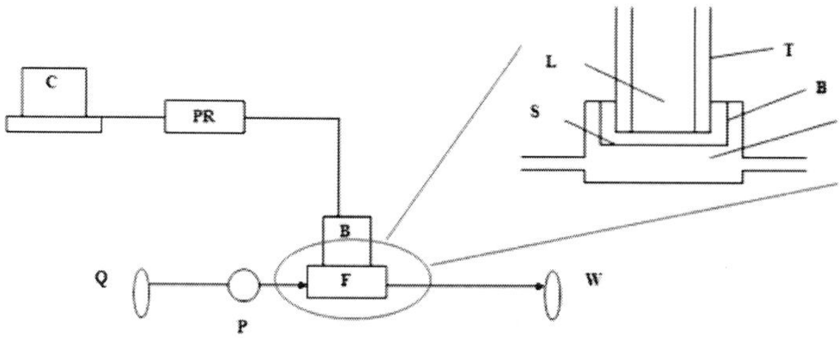

Figure 4.2 Schematic of the carbon utilization biosensor with a detailed diagram of the biosensor probe. The biosensor probe tip was inserted into the flow chamber. Water, waste, or medium streams were kept at a constant flow rate, controlled by a peristaltic pump. C; computer; PR: photomultiplier and recorder; B: biosensor; F: flow chamber; W: waste tank; P: peristaltic pump; Q: sample; L: light pipe; T: tubing with a glass glued inside of it; B: beads chamber; S: porous stainless steel plate. From Hartano et al. (2008) [13].

lac-luxCDABE transcriptional fusion and generates bioluminescence when sufficient energy is available for metabolism. This system was developed to monitor bioavailable carbon utilization by the recombinant food-borne pathogen [13]. In the constructed system, resting cells were immobilized in alginate beads using a strontium cross-linker. The beads were placed in a cylindrical chamber behind a stainless steel wire mesh and interfaced with a Hamamatsu PMT module. The biosensor was placed in a light-tight enclosure with inlet and outlet fittings, allowing sample flow (Fig. 4.2). The study indicated that an increase in bioluminescence was directly correlated with an increase in carbon concentration. The biosensor provides insight into pathogen maintenance and growth potential, which could lead to the development of alternative strategies for reducing contamination and enhanced safety measures.

4.4 Photomultipliers

Due to the necessity to generate very sensitive bioluminescent biosensor systems, most light throughput, regardless of hardware, requires augmentation of the light signal. In this regard, PMTs

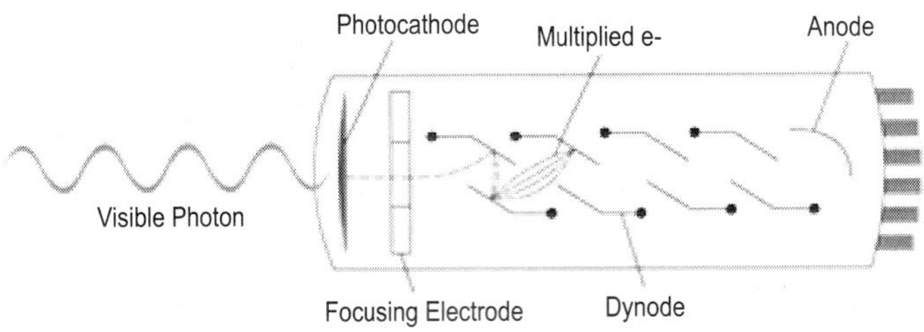

Figure 4.3 Simple schematic of a photomultiplier tube. Adapted from Finney (2007) [16].

are used to detect low-energy photons (ultraviolet-visible [UV-Vis] range), high-energy photons (X-rays and gamma rays), and even ionizing particles via scintillators. Attributes, including a large sensing area, fast response, and high gain/low noise, make photomultipliers attractive in bioluminescence detection [14].

A photomultiplier operates by carrying an electric charge proportional to the amount of light it receives. In this sense, it is dependent on the photoelectric effect. A metallic plate (dynode) will release electrons when it is struck by corresponding photons (Fig. 4.3).

The number of electrons released is dependent upon the intensity and frequency of the light [15]. The key aspect and namesake of this device is to multiply the signal. Thus, via repetitive metallic plate stages (typical range of 6–12), the liberated electrons from the initial plate accelerate toward downstream plates, where more electrons are released by colliding with said downstream plates. This system stands to have a tremendous gain due to the multiplicative factor. It is this characteristic that makes the use of PMTs in bioluminescent reporters so attractive. Figure 4.4 shows multiple dynode configurations that lead to the propagation of the photoelectric signal.

Research studies have used liquid light guides in tandem with PMTs. As stated previously, their ability to efficiently transfer light and multiply the signal allows these systems to be highly sensitive. In 1998, Elasri and Miller used a biosensor by constructing a plasmid containing a fusion of the *recA* promoter of *Pseudomonas aeruginosa* to the *luxCDABE* operon of *Vibrio fischeri* [7]. Using a

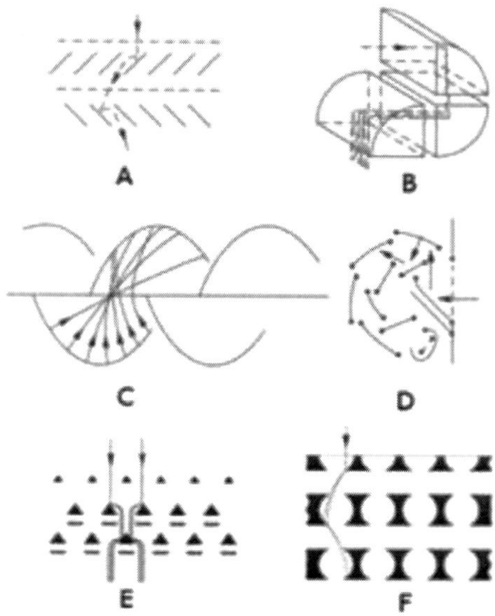

Figure 4.4 Varying dynode configurations with photomultipliers. A: venetian blind; B: box; C: linear focusing; D: circular cage; E: mesh; F: foil. From Flyckt and Marmonier (2002) [14].

liquid light guide and a photomultiplier setup, researchers were able to examine bacterial bioluminescence in response to UVC exposure of *P. aeruginosa* RM4440. This was one of the first studies utilizing a bioluminescent reporter in order to detect bioluminescent changes in response to UV-radiation-mediated DNA damage. Elasri and Miller found that the bioluminescence response was rapid and directly proportional to the UV dose.

Often, the ability to sense and measure compounds serves as a great instrument for regulatory agencies and/or industrial facilities. Detection techniques, such as liquid chromatography, can be expensive and time consuming. Advantages of bioluminescent biosensors are that they allow the identification of compounds at low concentrations, they are inexpensive, andm most importantly, they are rapid in performance. Thouand et al. (2003) made a bioluminescent biosensor that responded when in the presence of tributylin (TBT). In this study, the researchers created a bioreporter, *E. coli* TBT3, which was constructed from the random insertion of

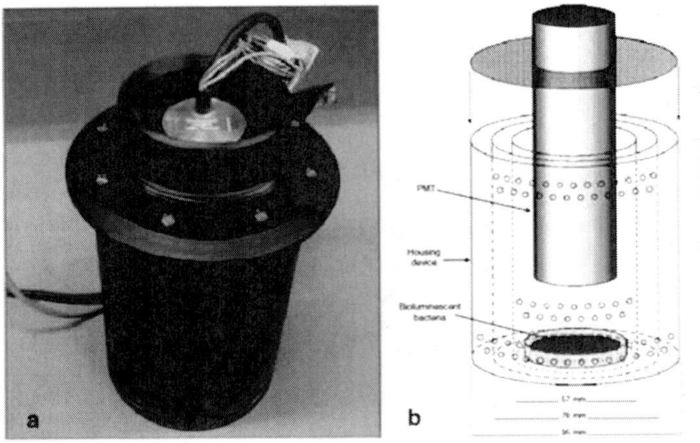

Figure 4.5 Biosensor designed for the online monitoring of chlorine dioxide gas disinfection. (a) Biosensor device housing a photomultiplier tube; (b) schematic representation of the biosensor designed for the online monitoring of chlorine dioxide gas disinfection. From Busto-Ramos et al. (2008) [5].

the *luxAB* genes of *Vibrio harveyi* into the chromosome of *E. coli* DH1 [17]. Using a system similar to that of Kelly et al. (1999), Thouand et al. were the first in reporting the online detection of TBT using a bioluminescent biosensor.

Comparatively, there are systems that may not require light transfer. In 2008, Busto-Ramos et al. used a similar system, but it lacked the liquid light guide [5]. In this study, Busto-Ramos et al. developed an online biosensor to monitor chlorine dioxide gas disinfection (Fig. 4.5). Researchers used a system that involved *Pseudomonas fluorescens* 5RL and a PMT that was directly mounted to the bioreactor chamber. The study utilized *P. fluorescens* 5RL as the bioreporter because of its ability to produce high light levels in the presence of salicylate.

To limit light entry but allow gas diffusion inside the chamber, the chamber was made of black Delrin material (polyoxymethylene), with three concentric cylinders of decreasing diameters, containing perforations in alternate positions [5]. Ultimately, Busto-Ramos et al. (2008) [5] used the system to demonstrate that greater ClO_2 gas concentrations directly correlated to the rate of decrease in bioluminescence. They were able to show that the biosensor had the ability to monitor and evaluate ClO_2 disinfection and other types of gas disinfection. However, the system was limited to only the kinetic analyses of bioluminescence due to the sampling restrictions and subsequent determination of viable bacterial colony-forming units (CFUs).

Biosensors that utilize fiber optic bundles, liquid light guides, and photomultipliers tend to be best suited for customization. They are widely adaptable to various applications and can be used to develop online, in situ systems. Additionally, each device comes in many different confirmations, which allows for even further customization. This plug-and-play style of setup is a major reason for their wide use in biosensors.

4.5 EMCCD Devices

Although some bioluminescent biosensors relay online or continuous data with such systems similar to those previously mentioned, others can be used to display visual data, or be programmed to take visual representations at set time points. Electron-multiplying charge-coupled devices (EMCCDs) have this purpose. An EMCCD camera, for example, operates similarly to a photomultiplier, with the exception of electrons being multiplied by impact ionization. This is similar to the avalanche diode mechanism utilized in photomultipliers. EMCCDs have notable quantum efficiency in excess of 70% with a wavelength range from 450 to 800 nm and as high as 90% from a 500–650 nm range [18]. With EMCCD cameras, incredible gain is achieved with minimal background noise. They also have the capability to detect a single photon [19].

Interestingly, calcium (Ca^{2+})-sensitive bioluminescent reporters undergoing bioluminescence resonance energy transfer (BRET) have been constructed with aequorin and green fluorescent protein (GFP) [18]. These hybrid proteins possess greater light intensity and stability than aequorin alone. Rogers et al. (2008) constructed a biosensor from HEK-293 cells, transfected to express GFP-apoaequorin, with an EMCCD camera mounted on bioluminescent and conventional microscopes as its transducer. Researchers demonstrated the use of a biosensor to detect receptor-driven Ca^{2+} oscillations in single cells via BRET. In this study, the biosensor was brighter and more sensitive via augmentation of the signal by BRET and the use of the EMCCD camera.

In 2012, Myer et al. (unpublished data) used an EMCCD camera to detect bacteriophage propagation within a bacterial lawn [20]. Using the EMCCD camera, *P. fluorescens* M3A bearing the salicylate-induced *nahRG*$_p$*-luxCDABE* cassette, and Phi-S1 bacteriophage, researchers detected bacteriophage plaque formation in within 6 hours of infection. This was significant when compared to the 8 hours required when examined visually for bacterial lawn formation, further demonstrating the powerful use of EMCCD cameras as a component of bioluminescent biosensors.

4.6 Luminometers

The biological and technological advancement of bioluminescent biosensors relies on the ability to maximize the bioluminescent signal, the ability to efficiently target the compound of choice, the ability to effectively transduce the signal, and adaptability. There are currently countless biosensors utilizing some form of a Lux, Luc, or Aequorin system. Even with advancements in sensor technology, as with the EMCCD or liquid light guides, luminometers have been a mainstay in bioluminescent research. Luminometers have practical applications because they can be both high throughput, as in microplate multilabel readers, or they can be small enough to be used on the benchtop.

Benchtop luminometers, or single-tube luminometers, are applicable in the realm of liquid culture-based biosensors rather than that of immobilized bioreporter biosensors. They rely on sampling the experimental reactor and manual insertion of the sample into the luminometer. Most luminometers detect light with photoresistors, photodiodes, or photomultipliers, and within the same system, the signal is quantified. This portable, benchtop, all-in-one unit makes the system quite attractive through ease of use alone.

There are obvious drawbacks to using a benchtop luminometer as both the transducer and the recorder of the biosensor system. Innately, benchtop models do not allow for online, autonomous, continuous data collection. Removal and manual manipulation of the sample can prove troublesome, accounting for increased penitential in human error. Additionally, the sample must be responsive and grown in liquid culture rather than be immobilized.

Firefly luciferase (Luc) is another luciferase commonly used in bioluminescent biosensors (Table 4.1). Because of the necessity to add the substrate, luciferin, real-time data still remains a challenge in utilizing this system. However, benefits of Luc include its fast reaction time and bright intensity or nature. Although much study has focused on bacterial bioreporters, eukaryotic bioreporters harboring components from bacterial sources has been difficult. Hollis et al. (2000) designed a biosensor using *Saccharomyces cerevisiae* with chromosomally expressed Luc from *Photinus pyralis*, a cuvette luminometer, and a method to apply luciferin before bioluminescence qualification [21]. In this study, the researchers tested various toxins to assess the level of toxicity on the cell. The light reaction of Luc relies on the availability of adenosine triphosphate (ATP), which is indicative of an actively metabolizing cell. Thus, this biosensor can directly report the health of the cell via the light reaction. Hollis et al. demonstrated the efficacy of their biosensor, while also determining that certain chemicals deemed nontoxic to bacterial biosensors are toxic in eukaryotic types [21].

The addition of automation and a small sample size to develop high-throughput luminometers have allowed scientists to accumulate large data sets encompassing different parameters to varying degrees. In these cases, microplate readers or multilabel readers are able to show rapid turnover of numerous samples. Such instruments utilize microplates, which contain a variable number of wells (e.g., 96-well plates) for individual sample analysis. The strength of this system is its ability to automate several functions, while quickly sampling. These systems can automate temperature controls, plate agitation, aeration, and luminometry/fluorometry. Though the operator must load individual samples (which is reminiscent of the benchtop luminometer), the operator is able to load different treatments of the same bioreporter, thus obtaining many data sets. This is different to the aforementioned liquid light guide and the PMT-based biosensor, in which the biosensor only developed data for the single parameter examined.

Kelly et al. (2003) worked with analyzing the kinetics of bacterial bioluminescence [22]. Kelly et al. used *P. fluorescens* HK44, a *lux*-based bioreporter developed by insertion of the *lux* gene cassette from *V. fischeri* into a naphthalene catabolic plasmid in *P. fluorescens* 5R [23]. *P. fluorescens* HK44 has the ability to sense naphthalene degradation. The bioreporter was transduced by a luminometer. In the study, the researchers used the biosensor in parallel with a biosensor utilizing a photosensor in order to examine the properties of the bacterial bioluminescence in terms of cellular energy limitations, growth substrate, and biomass concentrations [22]. This study was a prime example of how the use of a luminometer provides the sole source of bioluminescent data and/or enhance data acquisition.

4.7 Microluminometers

Although dogmatic and functional, the application of fiber optics or light guides, linked to PMTs and terminating at any number of optical transducers, requires that the system be somewhat large in size, necessitating greater power constraints and limiting portability. In many aspects, the use of biosensors outside of the laboratory setting is important for acquiring environmental data. The requirement for small sample volumes and remote detection implores one to examine other options. Thus, the previous systems are not best suited to miniaturization and portability.

Within the last decade, advances have allowed the formation of a system that bypasses the need for large detecting equipment and large sample volumes. These systems integrate the bioreporter to a circuit that can detect, process, and report low levels of bioluminescence. The bioluminescent bioreporter integrated circuit (BBIC) is such a biosensor that couples all the actions of the sensor into one, small, low-power device [24]. The concept requires that a bioluminescent bioreporter be immobilized onto the chip and the chip be bioresistant and biocompatible as well as perform all the necessary functions [8]. The advantage here is that though most biosensors have separate systems that must be linked and interfaced, the BBIC allows all the compatible functions to be bridged onto one device.

There are many different methods for the incorporation of the bioreporter into the integrated circuit. The cells can be placed behind a porous membrane or encapsulated into natural or synthetic polymeric substances. Encapsulation into polymeric substances is beneficial in that the cells encompass many protective and growth attributes. Encapsulation protects the cells from the environment and toxic substances. In addition, it provides the cells with the moisture and nutrients required to sustain a healthy and responsive culture. Usually, the application of substances to the bioreporter requires mixing the cell culture with a liquid polymer [10], micropipetting the mixture onto the integrated circuit, and allowing the resulting application to cure. One can also create microbeads by passing the mixture through a nebulizer, spraying onto a polymerizing agent, and then applying to the circuit [25]. Many different protocols exist, but common ones include the use of agar/agarose (1%–5%), carrageenan (2%), alginate (1%–8%), polyurethane-polycarbomyl sulfonate (PCS), polyacrylamide, poly(vinyl alcohol), and sol-gels.

As a whole, the BBIC measures bioluminescence from the bioreporter, digitizes the signal, and transmits the data to a data receiver [26]. Though the circuit cannot detect individual photons, which is indicative of PMTs, EMCCDs, and silicon photodetectors (see Section 4.9), it can detect low-level bioluminescence because the bioluminescent reaction is a relatively long-lived reaction [26]. The circuit includes voltage amplifiers and current-to-frequency converters to allow for signal amplification and data transmittance with subsequent recording. Figure 4.6 displays a layout and a simple schematic of a BBIC [27].

A study by Ripp et al. (2003) used a BBIC to detect volatile organic compounds (VOCs). The biosensor used the bioreporter *Pseudomonas putida* UT93. The reporter had the *luxCDABE* gene cassette from *V. fischeri* fused with the *p*-cymene-/*p*-cumate-inducible promoter, which was derived from a *P. putida* F1 *cym* operon [28]. The BBIC was able to ultimately detect volatile

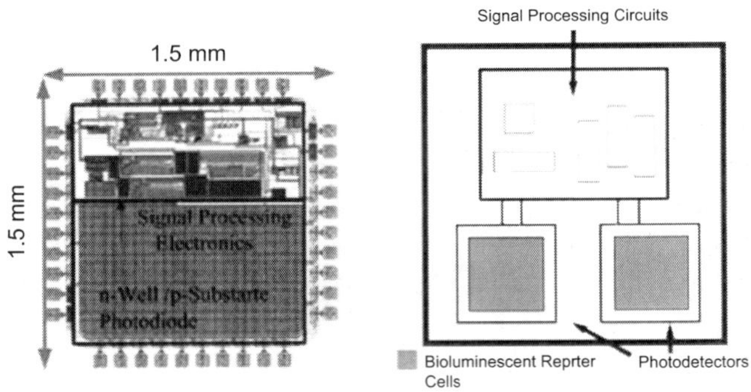

Figure 4.6 Layout and simple schematic of a BBIC. Adapted from Islam et al. (2007) [27].

metabolites (*p*-cymene) from *Penicillium roqueforti* within 3.5 h of exposure to the organism. This study demonstrated the application of the BBIC in parallel with its bioreporter. It also demonstrated the ability to indirectly detect microbes via bioluminescence reporting.

The BBIC has also been produced into a handheld wand format, applicable for field monitoring [28]. This device is operated off an internal lithium ion watch battery and engineered to completely encase a portable biosensor. Like the aforementioned BBIC, this device uses is able to sense the signal from the bioreporter cells, transduce the signal, and submit data. Figure 4.7 displays an example of the BBIC handheld wand.

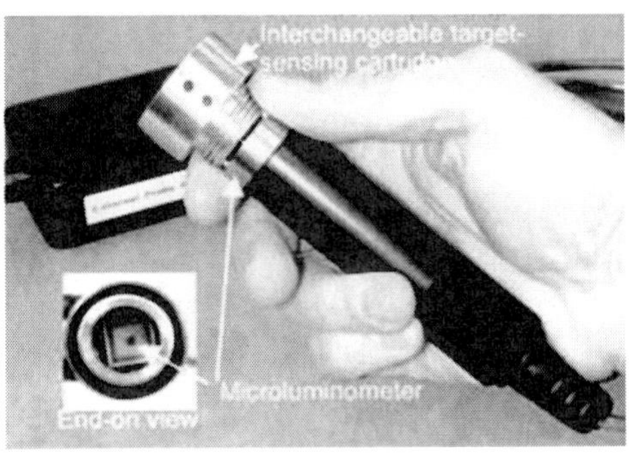

Figure 4.7 Example of a BBIC-powered microluminometer transducer in handheld wand format. From Ripp (2006) [30].

Interestingly, Eltoukhy et al. (2006) developed a microluminometer that contained a 128-channel array [29] capable of sensing individual and multiple bioreporters simultaneously. This makes it a very attractive system that can be applied in ascertaining multiple chemical analyses. Many different bioreporters, using different analyte–reporter combinations, can be utilized with potential for numerous combinations.

4.8 ATP Bioluminescence

Most of the above-mentioned literature has focused on biosensors with whole-cell signal attributes. These types of biosensors are usually rooted in the sensing of toxicity and resultant decrease in bioluminescence and/or the sensing of specific target analytes, thus influencing the bioluminescent response by induction of bioluminescence. Though their applications are broad and target specific, their setup and response time can be extensive. Additionally, with limited exception [28], these types of biosensors are not optimal at sensing the existence of other microbes. For airborne microbes, identification is conventionally laborious due to collection, culturing, and long incubation time. The BBIC system mentioned above has addressed this issue, but other systems have been developed in order to facilitate the detection of airborne microbes.

ATP bioluminescence is based on the concept that the D-luciferin-luciferase reagent from *P. pyralis* is ATP dependent. In the simplest of forms, ATP is used to fuel the luminescent reaction. Thus, with ATP being the most important biological fuel in living organisms, it is justifiable in determining that the detection of ATP from living organisms would be a sufficient indirect tool for the detection of living organisms. Indeed, there have been many studies that have addressed the issue of ATP and monitoring of biological and environmental systems. For example, the growth of bacteria has been monitored in real time by the measurement of bioluminescence [31].

Lee et al. (2008) developed a system that encompasses aspects of both portability and ATP bioluminescence (Fig. 4.8). In the study, the researchers developed a real-time detection system to

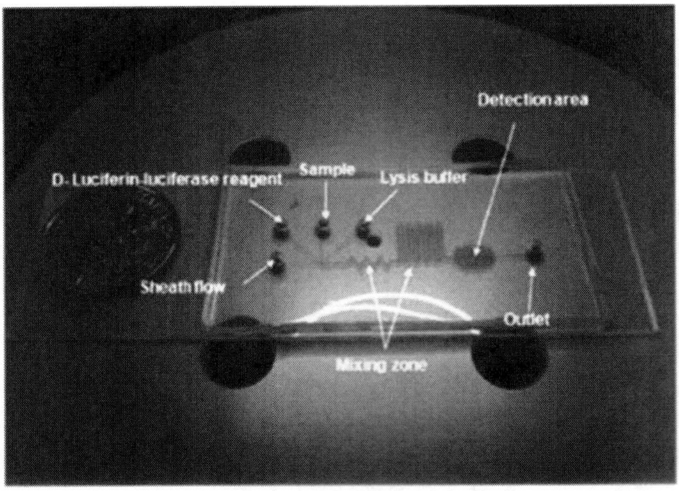

Figure 4.8 Photograph of the ATP bioluminescent microfluidic chip. From Lee et al. (2008) [32].

measure ATP extracted from airborne microbes [32]. Additionally, it was developed to be functional on a microfluidic chip. The chip allows the entire process to be embedded in one location for rapid, portable results. The captured airborne microbes are dissolved in the microfludic channel by a lysis buffer and the ATP is then extracted. This occurs all at once within the channel. Along with these actions, the D-luciferin-luciferase reagents are added and then the entire flow is mixed within the channel.

The released ATP reaction with the D-luciferin-luciferase reagents allows for the bioluminescent signal to be detected and the flow-through to be released through an outlet. Lee et al. (2008) were able to use their microfluidic chip to detect airborne microbes *Bacillus subtilis* and *E. coli* JM110 within 10 min [32].

This design was indicative of the methods that can lead to efficient, portable biosensors capable of sensing a variety of organisms. Portable biosensors and/or microluminometers are necessary and positive advancements in biosensor technology out of the laboratory setting and toward field monitoring. Their convenient size and continuing ability to sense a variety of specific targets make them attractive instruments.

4.9 Silicon Photomultipliers

Though their namesake suggests a relation to traditional photomultipliers, silicon photomultipliers (SPMs) are characteristically distinct from the traditional version. The traditional photomultiplier has been effective and accurate in photon-counting applications with regard to bacterial bioreporters. Its advantage is a low dark count and low threshold detection limit [33]. However, limitations and detractions in using this style of photomultiplier do exist. They use a great deal of energy, are expensive, and lack a dynamic linear range [34].

Photon-counting devices that are based on silicon Geiger-mode avalanche photodiodes (APDs) are recent additions to the photon-counting detector repertoire. These devices not only stand alone but also work in combination or addition with existing software and devices or are made into an array to develop highly sensitive equipment. SPMs utilize this Geiger-mode photodiode array. The SPM uses an array of microcells that counts photons and each microcell has a Geiger-mode photodiode. This array is then joined to produce a single combined signal. More specifically, each microcell is connected in parallel, which enables the device to output a signal that is proportional to the number of photons that have entered the device [35].

SPMs are similar to traditional photomultipliers, in that both have high gain and are sensitive to the resolution of a single photon. Yet, as stated previously, additional attributes make them distinct from said photomultipliers. SPMs are solid-state devices, which are significantly smaller and more compact than other photomultipliers. Also, their compact size allows for ease in portability and detection capability using small sample volumes, enabling applications in field monitoring. Unlike other photomultipliers, SPMs have a low operation voltage, which again can be beneficial in portability. Finally, these instruments are not affected by magnetic fields nor are they damaged by light overexposure, which are both common disadvantages of using traditional photomultipliers [35].

SPMs are not without their own challenges. In most of the devices, though high gain, the photon counts are usually lower than those of a PMT under similar conditions. This is attributed to the active area of the comparable devices, of which the active area of the PMT can be up to 50 times larger than that of the SPM [34].

Li et al. (2012) demonstrated this active area discrepancy by examining the differences between two biosensors, PMTs and SPMs [34]. Each biosensor had the bioluminescent strain *P. fluorescens* 5RL as its bioreporter component, which emits a strong bioluminescent response, and was key to detecting and examining higher levels of bioluminescence in the study. One of the main outcomes of their research was the discovery of differing linear relationships. With varying concentrations of the *P. fluorescens* 5RL *luxCDABE* cassette inducer salicylate, the calibration curves showed that though the PMT was more sensitive, concentrations greater than 500 ppb did not indicate a linear relationship. On the other hand, the SPM showed a linear relationship for the entire range of salicylate concentrations, up to 5000 ppb [34], demonstrating that though the active area may be smaller in the SPM, it maintains uniform linearity throughout its response profile. Overall, even though the PMT has a higher detection limit, the linearity of its detection, small solid-state size, low sample volume, and low operation voltage make the SPM attractive enough for use, especially in the context away from the benchtop.

4.10 Environmental Biosensor Applications

Most research follows one of two paths, the route of continued research or the route of applied research. Most applications of laboratory findings need to be examined outside of the laboratory context. Microbial whole-cell bioluminescent biosensors are no different. The trend of bioluminescent biosensors in today's research is the focus of miniaturizing or developing online sensors to analyze environmental systems. Examples of miniaturization for studies in field monitoring have already been mentioned. The use of BBIC and BBIC-powered handheld wands to detect a number of targets, dependent on the organism and its construction, has been studied and is currently in use [8, 28]. It is a powerful tool that uses low sample volume and low voltage to quickly detect a bioluminescent signal. Yet, its high detection limit is more a function of greater integration times rather than the device itself [34]. Additionally, SPMs have been used for the same reasons: their low voltage, small size, photon number resolution, and linear response [35]. Microfluidic ATP bioluminescent biosensors, which were discussed earlier, are further examples of bioluminescent biosensors with environmental applications [32].

Municipal sewage and industrial effluents are commonly laden with heavy metals and are usually treated with activated sludge for their removal in wastewater treatment systems [36]. It is often necessary to examine the system and the viability and efficacy of the activated sludge in order to assess the toxicity of the flow-through caused by residual heavy metals. Kelly et al. (1999) developed an online system to monitor toxicity in wastewater treatment systems. These researchers developed the bioluminescent reporter Shk1 from activated sludge bacteria via mating of *E. coli* containing a Tn4431 *lux* transposon [36]. This organism was identified via 16S rDNA sequencing to be *P. fluorescens.*

The toxicity-monitoring biosensor system for application in the biological wastewater system was developed to be an online, continuous sampling scheme. Figure 4.9 shows a schematic of the continuous sample system to measure the response of the Shk1 bioluminescent reporter to toxicants [12]. Hydroquinone was used as the toxicant because of its common relationship with industrial effluents. The system consisted of a chemostat that was responsible for the production of the bioreporter. This was constantly fed into individual testing cells that were light-sequestered. The control cell was fed simulated wastewater, while the experimental cell was fed the same simulated wastewater but was also fed increments of hydroquinone. A liquid light guide relayed the bioluminescent response to a photomultiplier and finally to a recorder. The experimental data were normalized to the control. The setup was able to monitor the toxicity of hydroquinone to Shk1 and displayed marked decreases in bioluminescence in response to its addition. This system is just one of many examples of how whole-cell bioluminescent biosensors can be applied, either for use to monitor environmental systems or in other field monitoring systems.

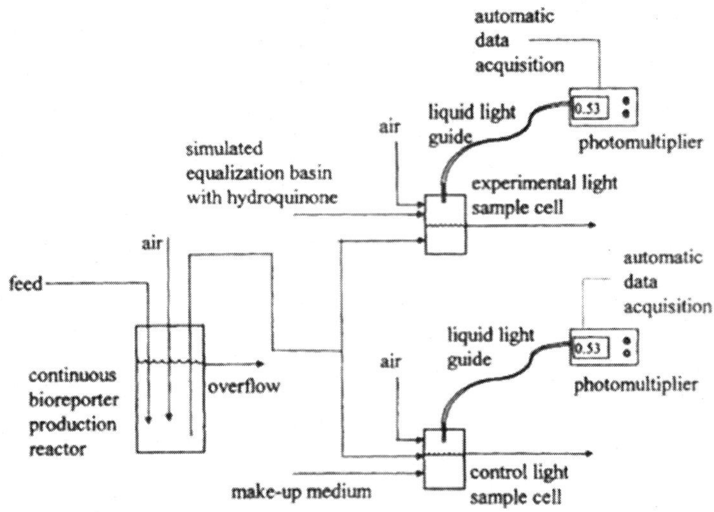

Figure 4.9 Schematic of the continuous sample system to measure the response of the Shk1 bioluminescent reporter to toxicants. From Kelly et al. (1999) [12].

4.11 Concluding Remarks

The field of bioluminescent biosensors has an unlimited potential due to the evolution in technology of each of its components. The whole-cell bioluminescent bioreporters can be applied to numerous targets, depending on the target of choice and how the organism is engineered. Not only is genetics an important component for alteration, but so too is the organism itself. There are different bioluminescent systems (Table 4.1) and each can be applied to numerous organisms: from bacteria, to archaea, to eukaryotes. Additionally, advances in technology constantly affect other components, including light guides, transducers, and photomultipliers. As the field of optics advances, so does one's ability to alter bioluminescent biosensors. Applications from this technology are far reaching and their ability to do so naturally is also appealing to the commercial and industrial sector. Though not a novel field, the design, construction, and application of bioluminescent biosensors have evolved greatly and still have a positive future with room for tremendous growth.

References

1. Ausubel, F. M., Brent, R., et al. (2002). *Short Protocols in Molecular Biology*, Wiley, Oxford.
2. de Lorenzo, V., Herrero, M., et al. (1990). Mini-Tn5 transposon derivatives for insertion mutagenesis, promoter probing, and chromosomal insertion of cloned DNA in gram-negative eubacteria. *J. Bacteriol.*, **172**(11), 6568–6572.
3. D'souza, S. F. (2001). Microbialbiosensors. *Biosens. Bioelectron.*, **16**(6), 337–353.
4. Elasri, M. O., Reid, T., et al. (2000). Response of a Pseudomonas aeruginosa biofilm community to DNA-damaging chemical agents. *FEMS Microbiol. Ecol.*, **33**(1), 21–25.
5. Busto-Ramos, M., Budzik, M., et al. (2008). Development of an online biosensor for in situ monitoring of chlorine dioxide gas disinfection efficacy. *Appl. Microbiol. Biotechnol.*, **78**(4), 573–580.
6. Heitzer, A., Applegate, B., et al. (1998). Physiological considerations of environmental applications of lux reporter fusions. *J. Microbiol. Methods*, **33**(1), 45–57.
7. Elasri, M. O., Miller, R. V. (1998). A Pseudomonas aeruginosa biosensor responds to exposure to ultraviolet radiation. *Appl. Microbiol. Biotechnol.*, **50**(4), 455–458.
8. Close, D., Ripp, S., et al. (2009). Reporter proteins in whole-cell optical bioreporter detection systems, biosensor integrations, and biosensing applications. *Sensors*, **9**(11), 9147–9174.
9. Webb, S. E., Miller, A. L. (2012). Aequorin-based genetic approaches to visualize Ca2+ signaling in developing animal systems. *Biochim. Biophys. Acta*, **1820**(8), 1160–1168.
10. Heitzer, A., Malachowsky, K., et al. (1994). Optical biosensor for environmental on-line monitoring of naphthalene and salicylate bioavailability with an immobilized bioluminescent catabolic reporter bacterium. *Appl. Environ. Microbiol.*, **60**(5), 1487–1494.
11. Newport, C. (2013). *Liquid Light Guides 2013*, http://www.newport.com/Liquid-Light-Guides/378731/1033/info.aspx#tab_Overview.

12. Kelly, C. J., Lajoie, C. A., et al. (1999). Bioluminescent reporter bacterium for toxicity monitoring in biological wastewater treatment systems. *Water Environ. Res.*, **71**(1), 31–35.

13. Hartano, L., Shroyer, M. L., et al. (2008). *Development of an Escherichia coli 0157:H7 Bioluminescent Reporter for Measuring Bioavailable Carbon*. American Society for Microbiology, Boston, MA, USA.

14. Flyckt, S.-O., Marmonier, C. (2002). *Photomultiplier Tubes Principles and Applications*. Photonis, Brive, France.

15. Brennan, T. (2001). *A Summary of the Design, Development, and Testing of Photomultiplier Tubes and Base Electronics for the BTeV Experiment*, Fermi National Accelerator Laboratory.

16. Finney, N. (2007). *Gamma-Ray Spectroscopy*, University of California, Berkeley, CA, USA.

17. Thouand, G., Horry, H., Durand, M. J., Picart, P., Bendriaa, L., Daniel, P., DuBow, M. S. (2003). Development of a biosensor for on-line detection of tributyltin with a recombinant bioluminescent Escherichia coli strain. *Appl. Microbiol. Biotechnol.* **62**(2–3), 218–225.

18. Rogers, K. L., Martin, J.-R., et al. (2008). Electron-multiplying charge-coupled detector-based bioluminescence recording of single-cell Ca2+. *J. Biomed. Opt.*, **13**(3), 031211.

19. Zhang, L., Neves, L., et al. (2009). A characterization of the single-photon sensitivity of an electron multiplying charge-coupled device. *J. Phys. B*, **42**, 114011.

20. Myer, P., Applegate, B., et al. (2012). *Use of Bioluminescence to Detect Bacteriophage Phi S1 Propagation in Pseudomonas fluorescens M3A*. Purdue Unviersity Department of Food Science, West Lafayette, IN, USA.

21. Hollis, R. P., Killham, K., et al. (2000). Design and application of a biosensor for monitoring toxicity of compounds to eukaryotes. *Appl. Environ. Microbiol.*, **66**(4), 1676–1679.

22. Kelly, C. J., Hsiung, C.-J., et al. (2003). Kinetic analysis of bacterial bioluminescence. *Biotechnol. Bioeng.*, **81**(3), 370–378.

23. King, J. M. H., DiGrazia, P. M., et al. (1990). Rapid, sensitive bioluminescent reporter technology for naphthalene exposure and biodegradation. *Science*, **249**, 778–781.

24. Simpson, M. L., Paulus, M. J., et al. (2008). Microluminometer chip and method to measure bioluminescence, UT–Battelle, USA, 7,371,538.

25. Kanasawud, P., Hjörleifsdottir, S., et al. (1989). Studies on immobilization of the thermophilic bacterium Thermus aquaticus YT-1 by

entrapment in various matrices. *Appl. Microbiol. Biotechnol.*, **31**(3), 228–233.

26. Simpson, M. L., Sayler, G. S., Applegate, B. M., Ripp, S., Nivens, D. E., Paulus, M. J., Jellison Jr, G. E. (1998). Bioluminescent-bioreporter integrated circuits form novel whole-cell biosensors. *Trends Biotechnol.*, **16**(8), 332–338.
27. Islam, S. K., Vijayaraghavan, R., et al. (2007). Integrated circuit biosensors using living whole-cell bioreporters. *Circ. Syst. I*, **54**(1), 89–98.
28. Ripp, S., Daumer, K. A., et al. (2003). Bioluminescent bioreporter integrated-circuit sensing of microbial volatile organic compounds. *J. Ind. Microbiol. Biotechnol.*, **30**(11), 636–642.
29. Eltoukhy, H., Salama, K., et al. (2006). A 0.18-um CMOS bioluminescence detection lab-on-chip. *IEEE J. Solid-State Circ.*, **41**(3), 651–662.
30. Ripp, S. (2006). Biosensor monitoring of the cleanroom environment. *CleanRooms Mag.*, **20**, 25–29.
31. Moore, G., Griffith, C. (2002). A comparison of traditional and recently developed methods for monitoring surface hygiene within the food industry: an industry trial. *Int. Environ. Health Res.*, **12**(4), 317–329.
32. Lee, S. J., Park, J. S., et al. (2008). A microfluidic ATP-bioluminescence sensor for the detection of airborne microbes. *Sens. Actuators B*, **132**(2), 443–448.
33. Jackson, C. (2007). Silicon photomultiplier detectors for low-light detection. *Photon. Spectra*, **41**(12), 64–68.
34. Li, H., Lopes, N., et al. (2012). Silicon photomultiplier (SPM) detection of low-level bioluminescence for the development of deployable whole-cell biosensors: possibilities and limitations. *Biosens. Bioelectron.*, **33**(1), 299–303.
35. Stewart, A. G., Wall, L., et al. (2009). Properties of silicon photon counting detectors and silicon photomultipliers. *J. Mod. Opt.*, **56**(2–3), 240–252.
36. Kelly, C. J., Tumsaroj, N., et al. (2004). Assessing wastewater metal toxicity with bacterial bioluminescence in a bench-scale wastewater treatment system. *Water Res.*, **38**(2), 423–431.

Chapter 5

Biosensor with Immobilized Cells and Lensless Imaging Detection

Luca Cevenini,[a] Aldo Roda,[a,b] and Elisa Michelini[a,b]

[a] *Istituto Nazionale di Biostrutture e Biosistemi, Roma, Italy*
[b] *Department of Chemistry "G. Ciamician," University of Bologna, Bologna, Italy*
elisa.michelini8@unibo.it

5.1 Introduction

Recent scientific literature demonstrates a growing interest in the development of miniaturized, portable analytical devices for on-field analysis. Such devices would be helpful in several fields, such as medical diagnostics, environmental toxicology, food quality assessment, and security applications for detecting chemical and biological threat agents [1]. Therefore the main challenge is to obtain robust, compact, low-cost devices able to perform rapid, ultrasensitive analyses without the need for equipped laboratories or skilled personnel. Moreover, the possibility to perform the simultaneous detection of multiple analytes with a single device is remarkably attractive.

Bioluminescent Microbial Biosensors: Design, Construction, and Implementation
Edited by Gérald Thouand and Robert S. Marks

ISBN 978-981-4613-65-1 (Hardcover), 978-981-4613-66-8 (eBook)
www.panstanford.com

Biosensors, thanks to their ability to exploit highly specific biomolecular recognition mechanisms, have shown to satisfy many of the analytical requirements related to on-field analysis, for example, sensitivity, selectivity, robustness, rapidity, reusability, and contained costs. Many examples of biosensor applications for onsite analysis have been recently reported, ranging from label-free surface plasmon resonance immunosensors to amperometric biosensors, aptamer–nanoparticle biosensors, and cell-based impedance sensors [2–4].

In particular, with the recent improvements in molecular biology techniques, cell-based biosensors are now emerging as sensitive analytical tools [5–7]. They are obtained by introducing into the cell a reporter protein, whose expression is regulated by a molecular recognition element (e.g., a receptor) and promoter sequences. The interaction of the recognition element with the analyte activates the expression of the reporter protein, which can be readily measured, for example, by fluorescence or bioluminescence (BL) detection, and is directly related to the bioavailable fraction of the analyte in the sample.

The development of portable cell-based biosensors is hampered by many unsolved issues. The main problem is that when analyzing real samples, living cells are in direct contact with the sample matrix, such as wastewaters or clinical samples, that may affect cell viability, thus causing a false uncontrolled analytical signal. Moreover the dynamic range is still limited and very often cell-based biosensors lack of sensitivity and selectivity. Another issue in cell-based biosensor fabrication is the need to integrate cells in a robust and portable analytical device, that is, immobilize them without affecting cell viability and sensing capacity. To this end different techniques have been exploited, including freeze-drying, the use of spores, and the immobilization in biocompatible polymers [8–11].

We envisaged the possibility of solving some of the pitfalls of whole-cell biosensors and turning them into efficient bioanalytical tools that could be implemented into portable devices for on-field applications.

Here we report the development of a portable biosensing device, relying on lensless contact imaging, composed of a disposable cartridge of immobilized BL whole-cell biosensors coupled with a lensless charge-coupled device (CCD) detector.

BL reporter gene technology was chosen thanks to the low background and high quantum yield of BL reactions (e.g., quantum yield for *Photinus pyralis* luciferase $\Phi = 0.44$) [12]. Besides, the availability of newly cloned thermostable BL reporter proteins with different emission properties prompts the development of multianalyte detection BL systems based on spectral resolution [13].

Yeast and bacterial whole-cell biosensors, expressing luciferases emitting at different wavelengths or requiring different BL substrates, were genetically engineered and integrated in the device to expand its multiplexing capability [14]. A new polymeric matrix based on natural and synthetic polymers (agarose, collagen, and poly[vinyl pyrrolidone] [PVP]) able to keep the immobilized cells alive was developed in order to obtain a ready-to-use biosensing cartridge that can be stored for up to 1 month without losing either its viability or the analyte's responsiveness.

5.2 Main Wished Features of the Device Based on Immobilized Biosensors and Lensless Imaging Detection

The immobilization of whole-cell biosensors into the new polymeric matrix allows the integration of living cells into a portable detection system that can be used outside laboratories even by nonskilled personnel.

Therefore, the first peculiarity of our system is portability and ease of use. In fact BL detection provides a simple readout and, differently from fluorescence, does not require an excitation light source or expensive instrumentation. In addition the use of improved BL proteins emitting at different wavelengths, with increased thermostability, and enhanced emission decay kinetic permits the analytical performance of whole-cell biosensors to improve. In particular the introduction of an additional BL reporter allows the correction of the analytical response according to cell viability, thus increasing the robustness of the bioassays. We previously reported the feasibility of using a combination of two or more BL proteins for obtaining biosensors with internal signal correction or for the simultaneous detection of multiple analytes [14]. This multiplexed biosensing platform containing whole-cell biosensors was used to measure the overall toxicity of a given sample as well as to obtain dose–response curves for heavy metals and to detect hormonal activity in clinical samples.

Since the assay requires just the sample and BL substrate addition a totally self-contained device could be set up just by including microfluidics for sample and reagent handling and user-friendly software.

Another significant advantage of this device is its high multiplexing capability, that is, the possibility to perform multianalyte analysis. Multiplex imaging detection can be in fact achieved by using a CCD camera and by (a) spatially resolving the light emitted by different luciferases, (b) sequentially measuring the light signals triggered by different substrates (e.g., D-luciferin and coelenterazine), and (c) spectrally resolving signals obtained from luciferases emitting at different wavelengths (e.g., luciferase mutants obtained from the same organism).

Furthermore, immobilized cells can be stored (at 4°C) for up to 1 month without losing viability or responsiveness [14]. This means that cells can be considered like a ready-to-use reagent and there is no need to culture cells in liquid media to revitalize them, thus simplifying the assay, while reducing assay cost and time.

5.3 Descriptions

5.3.1 *Recombinant Strains*

Different BL microbial bioreporters based on the use of *Saccharomyces cerevisiae* and *Escherichia coli* cells were developed. Both strains were genetically engineered by introducing into the same cell two BL reporter genes whose expression is either constitutive (viability control) or regulated by the activation of a specific biomolecular recognition element (e.g., a receptor). In particular the yeast bioreporter is based on a *S. cerevisiae* strain expressing a green-emitting luciferase (*P. pyralis* wild type) in response to

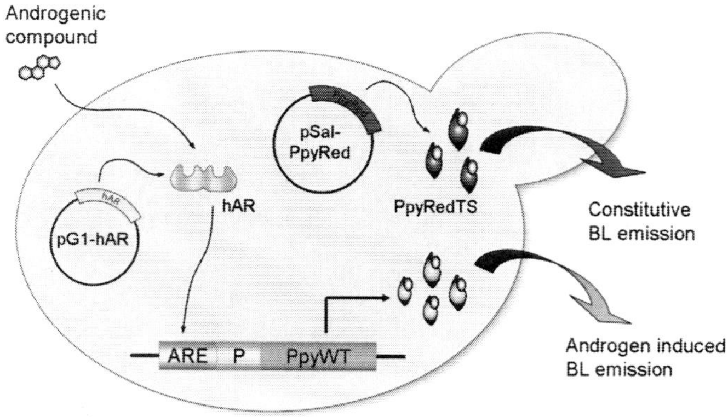

Figure 5.1 Schematic representation of the yeast androgen bioreporter with an internal viability control.

the activation of the human androgen receptor (hAR) and a red-emitting luciferase (the *P. pyralis* mutant Ppy RE8) used as an internal viability control. Figure 5.1 shows a schematic drawing of the androgen-responsive yeast bioreporter.

This strain was obtained by introduction of a red-emitting luciferase as the reporter gene for viability control into the recombinant BL yeast strain for androgen activity measurement previously described in Ref. [15]. Briefly, the parent strain was engineered by introducing an integrative vector in which *P. pyralis* wild-type luciferase (PpyWT) is under the regulation of the androgen-responsive element (ARE) sequences. Cells were also transformed with a plasmid for hAR constitutive expression. The red thermostable mutant luciferase Ppy RE8 [16] was cloned into the vector pSal1 [17], and the plasmid pSal1-Ppy RE8 was transformed into yeast using the lithium acetate method [18]. The recombinant strain was grown in a selective synthetic complete (sc) medium containing a 6.7 g/L yeast nitrogen base without amino acids; a 1.4 g/L yeast synthetic drop-out media supplement; and 40 mL/L of a 50% (w/v) D-glucose without uracil, tryptophan, and leucine.

The copper sensor strain *E. coli* MC1061 (pSLcueR/pDNPcop-Aluc) harbors the gene for the regulatory protein *cue*R and the promoter of copA, driving the expression of PpyWT luciferase [19]. It was engineered to express Ppy RE8 luciferase under the regulation of a promoter inducible by anhydrotetracyclin (ATc). Briefly, the cDNA encoding for Ppy RE8 was inserted into the multiple cloning site of the pPROTet.E133 vector by blunt ligation. Cells were then made chemically competent with the one-step transformation and storage solution method [20]. The plasmid pPROTet.E-PpyRE8, carrying chloramphenicol resistance, was transformed into this strain to obtain the microbial bioreporter with internal control. This strain was routinely grown in a selective Luria-Bertani (LB) medium containing 100 μg/mL ampicillin, 12.5 mg/mL ATc, and 25 μg/mL chloramphenicol.

5.3.2 *Immobilization of the Bioreporters*

Yeast and bacterial cells were immobilized into a polymeric matrix composed of agarose 0.5%, PVP 1%, and collagen solution 0.05% final concentration. Agarose and the PVP solution were prepared twofold concentrated in distilled water and sterilized by autoclaving at 121°C for 15 min.

Yeast and bacterial biosensors were cultured in a selective medium to midlog phase (optical density at 600 nm [OD600] $\approx$ 0.6 for bacteria and 0.8 for yeast cells). Then cell cultures were mixed 1:1 (v/v) with a matrix solution. A volume of 30 μL of the cell–matrix mixture was poured into the wells of the cartridge at room temperature, hardened for 30 min, and then kept at 4°C for long-term storage.

The cell-to-matrix ratio was optimized to improve the analytical performance of immobilized cells, in terms of both cell viability and sensing capability. In terms of reproducibility and BL response, the best results were obtained by immobilizing cells at 50% (v/v) with a mixture of agarose, PVP, and collagen.

The effect of matrix thickness on substrate diffusion and/or oxygen limitation was also investigated using matrix layers with a thickness of 1 mm, 2 mm, and 3 mm. As expected the matrix thickness was critical for BL substrate diffusion in cell–matrix layers and also affects the bioreporters' sensitivity. In a 3 mm cell–matrix layer, 60% of the BL signal arose from the uppermost 1 mm layer 30 s after substrate addition, while the middle and lower layers only

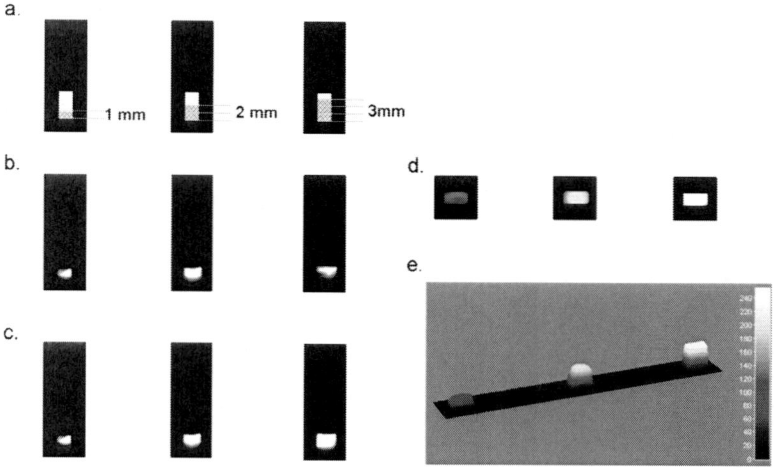

Figure 5.2 BL *E. coli* cells were immobilized in different cell–matrix layers inside blackened 1 mL cuvettes to investigate the effect of matrix thickness on substrate diffusion and/or oxygen limitation. BL signal acquisitions were performed with a NighOwl luminograph. (a) Schematic representation of matrix layers with a thickness of 1 mm, 2 mm, and 3 mm into blackened 1 mL plastic cuvettes. (b) BL emissions acquired 30 s after substrate addition. (c) BL emissions acquired 5 min after substrate addition. (d) BL emissions imaged from the top of the cuvettes. (e) An image from the top analyzed with an interactive ImageJ 3D surface plot.

contributed to 30% and 10% of the total BL emission, respectively (Fig. 5.2b; image on the right corresponding to a 3 mm thick layer).

In a 3 mm thick layer the BL signal intensity was limited by substrate diffusion, and it took around 5 min for the BL signal to be almost uniformly distributed across the whole cell–matrix layer (Fig. 5.2b and Fig. 5.2c; images on the right corresponding to a 3 mm thick layer). Although 1 and 2 mm cell–matrix layers provided faster diffusion of substrates, the total BL emissions were lower (Fig. 5.2d and Fig. 5.2e). According to our optimization experiment we decided to use 3 mm thick cell–matrix layers, corresponding to 30 μL of the cell–matrix mixture in a 384-well microplate.

The viability of immobilized cells stored at 4°C was monitored for 1 month, both in terms of BL emission (responsiveness) and as the percentage of living cells. Immobilized cells were suspended in

1 mL of growth medium by heating at 40°C in a water bath and being diluted 10^6-fold. A volume of 100 μL of bacterial or yeast suspension was spread on three replicas of selective agar plates, and the colony-forming units (CFUs) per well were counted after 24 h or 48 h incubation, respectively. The polymeric matrix was able to keep cells alive in ready-to-use cartridges for up to 35 days at 4°C.

5.3.3 *Device*

One of the most challenging issues for the development of portable cell-based biosensors for onsite analysis is the integration of the biological sensing element with a suitable detection system. We developed a BL-based multiplexed biosensing platform that includes a disposable microwell array cartridge, with immobilized whole-cell bioreporters, placed in contact with a CCD sensor through a fiber optic taper (Fig. 5.3).

The CCD camera used for the development of the portable device was selected thanks to its simple structure, only 400 g of weight, small dimensions, and the possibility of cooling the sensor with a built-in double Peltier chamber. The camera is a sensitive low-light Sony MZ-2PRO that is 83 mm wide, 83 mm deep, and 52 mm high (San Marco Optical Centre, Pordenone, Italy) with a 10.2 × 8.3 mm sensor at 1.4 megapixel (MP) resolution and a 16-bit analogue-to-digital converter. A fiber optic taper consisting of a mosaic of coherent optical fibers was used to enlarge the CCD light-sensitive area from 85 mm^2 to 200 mm^2.

An aluminum-forged adaptor with a neoprene O-ring was used to maintain the CCD sensor in contact with the taper and to provide thermal insulation. A customized 3 × 4 well array cartridge was fabricated by cutting a commercial black 384-well polystyrene microtiter plate with a transparent bottom (Labsystems, Helsinki, Finland). The cartridge with immobilized cells can be easily placed on the fiber optic taper, which allows for the detection of the light from each well of the cell array simultaneously. In addition, plastic filters can be inserted between the cartridge and the optical taper surface to obtain spectral resolution of light emissions at different wavelengths. In particular, a set of filters ranging from 500 nm to 700 nm was designed according to the red, green,

Figure 5.3 Picture of the device components. (a) Sony MZ-2PRO CCD-cooled camera with the fiber optic taper. (b) Clear-bottom cartridges with immobilized cells. (c) Picture of the assembled device with interchangeable cartridges in which different whole-cell bioreporters have been immobilized. (d) Overview of the portable device connected with a laptop via a USB cable ready for the analysis.

and blue (RGB) color values obtained with the spectra application (www.efg2.com/Lab/ScienceAndEngineering/Spectra.htm), printed on polyester (poly[ethylene terephthalate]) transparency films and characterized with the Cary Eclipse fluorescence spectrophotometer (Varian, Palo Alto, CA), by measuring the transmittance from 400 nm to 800 nm. The baseline was corrected by measuring the transmittance of nonprinted transparency film. According to the luciferases emission spectra, the best results were obtained with PF500 (plastic filter for green emission) (R:0, G:255, B:146) and PF640 (plastic filter for red emission) (R:255, G:33, B:0) filters printed with HP Color LaserJet 3600 on transparency films (3M, St. Paul, MN). Filters were printed as circles with a diameter of 3 mm to fit into 384-microwell format in order to be directly coupled to the cell cartridge.

5.3.4 *Electronic and Data Process*

One of the major advantages of using the CCD camera as a detector relies on the possibility of cooling the sensor with a Peltier chamber down to approximately to $-20°C$, allowing a significant decrease of the CCD dark current generated by thermal energy. It has been in fact demonstrated that the dark-current shot noise, whose contribution is often the major source of noise, is cut approximately in half for every $6°C$ reduction in chip temperature [21]. The variation due to dark-current shot noise is described by the equation

$$\sigma_{\mathrm{d}}^2 = 2qI_{\mathrm{d}}\Delta f \tag{5.1}$$

where σ_{d}^2 = variance of shot noise, q = charge of an electron, I_{d} = amount of dark current, and Δf = electronic bandwidth of the sensing system.

The thermoelectrical cooling system, although requiring sealing of the CCD sensor and electronics to prevent the fogging phenomenon and a larger power supply, together with the high-quantum efficiency of the Sony ICX285 CCD sensor (higher than 50% in the 420–680 nm range) allows the detection of very weak luminescent emissions with high reproducibility.

In fact we have previously reported that the background of the CCD is significantly reduced in going from $25°C$ to $-10°C$. Interestingly the standard deviation (SD) of the background signal measured at $-10°C$ is about 1 order of magnitude lower than that obtained at $25°C$ [22].

The device was connected with a laptop, and the CCD-Cap software V2.3.1 was used for parameter setup and images acquisition. Images were acquired using a 30 s exposure time and analyzed with both the NightOwl WinLight 32 and the ImageJ 1.42q software. The BL signal of each well was measured by integrating the photon emission over the well area. The background signal was subtracted from all the wells. Dose–response curves were corrected by reporting for each measurement the ratio of the analyte-specific light signal to the internal control signal (BL emission induced by analyte/ BL emission of the constitutive expressed luciferase [BL_S/BL_C]) against the log [analyte]. GraphPad Prism software v5.02 (Software, Inc., San Diego, CA) was used to fit the data for dose-response curves. The equation for the best-fit line, produced by the program, was used to calculate the amount of analyte present in the sample. The limit of detection (LOD) was defined as the concentration of copper giving a BL_S/BL_C value corresponding to that of the blank plus three times its SD.

5.4 Proof of Concept

The first proof-of-concept experiments were performed with a bacterial bioreporter immobilized for 2 weeks in the matrix and treated with dimethyl sulphoxide (DMSO), used as a model killing agent, to investigate the suitability of the device as a first-level toxicity screen. Figure 5.4 shows the BL intensity decrease at increasing DMSO concentrations.

Assay reproducibility was evaluated at a fixed concentration of DMSO (10% v/v); an intra-assay variability of 11% and an inter-assay variability of 17% ($n = 6$) were obtained. The assessment of the effects of compounds on cell viability is commonly used as a predictor of potential in vivo toxic effects in animals and humans. Therefore, the availability of low-cost devices that could be rapidly used onsite is extremely important [23].

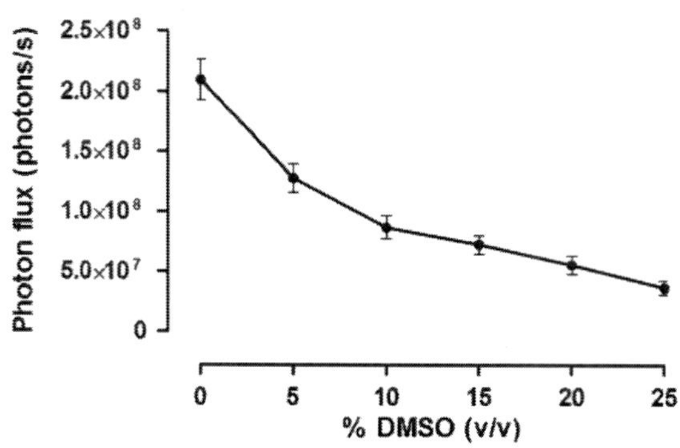

Figure 5.4 Toxicity curve for DMSO with immobilized *E. coli* bioreporter cells expressing a thermostable luciferase. Images were acquired using a 30 s exposure time, and the BL signal of each well was measured by integrating the photon emission over the well area.

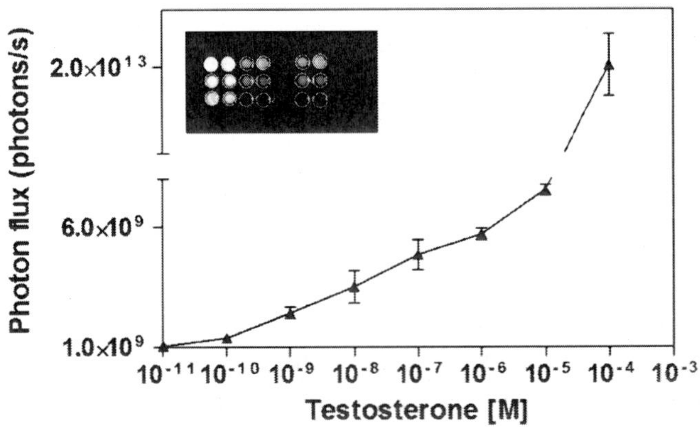

Figure 5.5 Corrected dose–response curve for testosterone obtained with the dual-color androgen-responsive strain immobilized in the device equipped with printed filters. Values are the mean ± standard deviations of triplicate samples. Inset: BL images of the cartridges with immobilized bioreporters incubated with different concentrations of testosterone.

To investigate the performance of the device dose–response curves, testosterone and copper ions were obtained using cartridges containing immobilized yeast bioreporters or a bacterial bioreporter strain, respectively.

A concentration–response curve for testosterone was obtained using the device equipped with filters, and an LOD of 0.5 nM was obtained (Fig. 5.5). This value is only 1 order of magnitude higher than that obtained with expensive standard benchtop instrumentation (LOD: 0.05 nM for testosterone) and customized high-transmission filters.

Then a comparison of dose–response curves for copper ions with the *E. coli* bioreporter obtained with the portable device and those obtained with a standard benchtop luminograph was performed. The resulting LOD of 6×10^{-6} M obtained with the device is slightly higher than that achieved with the same recombinant cells using the luminograph (1×10^{-6} M) (Fig. 5.6). However, considering the small dimension and the ease of use of this portable prototype, these results encourage us to improve the device and confirm the feasibility of low-cost multiplex BL portable devices for onsite applications.

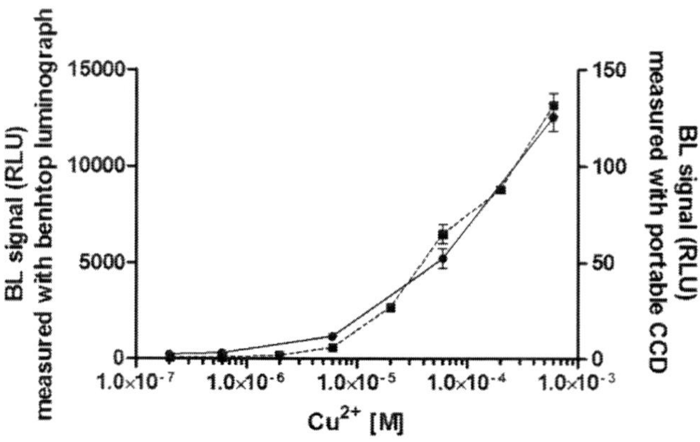

Figure 5.6 Corrected dose–response curve for copper ions obtained with the *E. coli* bioreporter immobilized in the device (■) and with conventional benchtop instrumentation (•).

Acknowledgments

The authors gratefully acknowledge the financial support provided by the Italian Ministry of University and Research (MIUR) with the project PRIN 2009MB4AYL and the project FIRB 2008, "Development of Bioremediation Procedures and Novel Detection Biotechnologies for Endocrine Disruptors" (prot. RBFR08SZTR). We are very grateful to Prof. Bruce Branchini (Connecticut College, New London, CT) for providing the cDNA encoding for Ppy RE8.

References

1. Aravanis, A. M., DeBusschere, B. D., Chruscinski, A. J., Gilchrist, K. H., Kobilka, B. K., Kovacs, G. T. (2001). A genetically engineered cell-based biosensor for functional classification of agents. *Biosens. Bioelectron.*, **16** (7–8), 571–577.
2. Carvalhal, R. F., Machado, D. S., Mendes, R. K., Almeida, A. L., Moreira, N. H., Piazetta, M. H., Gobbi, A. L., Kubota, L. T. (2010). Development of a disposable amperometric biosensor for salicylate based on a plastic electrochemical microcell. *Biosens. Bioelectron.*, **25**(10), 2200–2204.

3. Curtis, T. M., Widder, M. W., Brennan, L. M., Schwager, S. J., van der Schalie, W. H., Fey, J., Salazar, N. (2009). A portable cell-based impedance sensor for toxicity testing of drinking water. *Lab Chip*, **9**(15), 2176–2183.
4. Men, D., Zhang, Z. P., Guo, Y. C., Zhu, D. H., Bi, L. J., Deng, J. Y., Cui, Z. Q., Wei, H. P., Zhang, X. E. (2010). An auto-biotinylatedbifunctional protein nanowire for ultra-sensitive molecular biosensing. *Biosens. Bioelectron.*, **26**(4), 1137–1141.
5. Struss, A., Pasini, P., Ensor, C. M., Raut, N., Daunert, S. (2010). Paper strip whole cell biosensors: a portable test for the semiquantitative detection of bacterial quorum signaling molecules. *Anal. Chem.*, **82**(11), 4457–4463.
6. Misawa, N., Mitsuno, H., Kanzaki, R., Takeuchi, S. (2010). Highly sensitive and selective odorant sensor using living cells expressing insect olfactory receptors. *Proc. Natl. Acad. Sci. U S A*, **107**(35), 15340–15344.
7. Michelini, E., Cevenini, L., Mezzanotte, L., Coppa, A., Roda, A. (2010). Cell-based assays: fuelling drug discovery. *Anal. Bioanal. Chem.*, **398**(1), 227–238.
8. Amoura, M., Nassif, N., Roux, C., Livage, J., Coradin, T. (2007). Sol-gel encapsulation of cells is not limited to silica: long-term viability of bacteria in alumina matrices. *Chem. Commun.*, **39**, 4015–4017.
9. Choi, S. H., Gu, M. B. (2002). A portable toxicity biosensor using freeze-dried recombinant bioluminescent bacteria. *Biosens. Bioelectron.*, **17**, 433–440.
10. Date, A., Pasini, P., Daunert, S. (2010. Integration of spore-based genetically engineered whole-cell sensing systems into portable centrifugal microfluidic platforms. *Anal. Bioanal. Chem.*, **398**(1), 349–356.
11. Michelini, E., Roda, A. (2012). Staying alive: new perspectives on cell immobilization for biosensing purposes. *Anal. Bioanal. Chem.*, **402**(5), 1785–1797.
12. Roda, A., Pasini, P., Mirasoli, M., Michelini, E., Guardigli, M. (2004). Biotechnological applications of bioluminescence and chemiluminescence. *Trends Biotechnol.*, **22**(6), 295–303.
13. Michelini, E., Cevenini, L., Mezzanotte, L., Ablamsky, D., Southworth, T., Branchini, B., Roda, A. (2008). Spectral-resolved gene technology for multiplexed bioluminescence and high-content screening. *Anal. Chem.*, **80**(1), 260–267.

14. Roda, A., Cevenini, L., Michelini, E., Branchini, B. R. (2011). A portable bioluminescence engineered cell-based biosensor for on-site applications. *Biosens. Bioelectron.*, **26**(8), 3647–3653.
15. Michelini, E., Leskinen, P., Virta, M., Karp, M., Roda, A. (2005). A new recombinant cell-based bioluminescent assay for sensitive androgen-like compound detection. *Biosens. Bioelectron.*, **20**(11), 2261–2267.
16. Branchini, B. R., Ablamsky, D. M., Davis, A. L., Southworth, T. L., Butler, B., Fan, F., Jathoul, A. P., Pule, M. A. (2010). Red-emitting luciferases for bioluminescence reporter and imaging applications. *Anal. Biochem.*, **396**(2), 290–297.
17. Mascorro-Gallardo, J. O., Covarrubias, A. A., Gaxiola, R. (1996). Construction of a CUP1 promoter-based vector to modulate gene expression in Saccharomyces cerevisiae. *Gene*, **172**(1), 169–170.
18. Gietz, R. D., Schiestl, R. H. (1995). Studies on the transformation of intact yeast cells by the LiAc/SS-DNA/PEG procedure. *Yeast*, **11**(4), 355–360.
19. Hakkila, K., Green, T., Leskinen, P., Ivask, A., Marks, R., Virta, M. (2004). Detection of bioavailable heavy metals in EILATox-Oregon samples using whole-cell luminescent bacterial sensors in suspension or immobilized onto fibre-optic tips. *J. Appl. Toxicol.*, **24**(5), 333–342.
20. Chung, C. T., Niemela, S. L., Miller, R. H. (1989). One-step preparation of competent Escherichia coli: transformation and storage of bacterial cells in the same solution. *Proc. Natl. Acad. Sci. U S A*, **86**(7), 2172–2175.
21. Christensen, D. A., Herron, J. N. (2009). Optical system design for biosensors based on CCD detection. *Methods Mol. Biol.*, **503**, 239–258.
22. Roda, A., Mirasoli, M., Dolci, L. S., Buragina, A., Bonvicini, F., Simoni, P., Guardigli, M. (2011). Portable device based on chemiluminescencelens-less imaging for personalized diagnostics through multiplex bioanalysis. *Anal. Chem.*, **83**(8), 3178–3185.
23. Riss, T. L., Moravec, R. A., Niles, A. L. (2011). Cytotoxicity testing: measuring viable cells, dead cells, and detecting mechanism of cell death. *Methods Mol. Biol.*, **740**, 103–114.

Chapter 6

A Biosensor with Genetically Modified Bacteria Immobilized on a Fiber and a Glass Slide

Marjolijn Woutersen,[a] Bram van der Gaag,[a] Jan Mink,[b] Robert S. Marks,[c] Bram Brouwer,[d,e] and Minne B. Heringa[a]

[a] *KWR Watercycle Research Institute, Groningenhaven 7, 3433 PE Nieuwegein, The Netherlands*
[b] *2M Sensors, Torenallee 20, 5617 BC Eindhoven, The Netherlands*
[c] *Department of Biotechnology Engineering, Ben-Gurion University of the Negev, Beer-Sheva 84105, Israel*
[d] *Vrije Universiteit Faculty of Earth and Life Sciences, Department of Animal Ecology, De Boelelaan 1085, 1081 HV Amsterdam, The Netherlands*
[e] *BioDectection Systems, Science Park 406, 1089 XH Amsterdam, The Netherlands*
bram.brouwer@bds.nl

6.1 Introduction

To prevent contamination peaks that reach drinking water production, inlet points for surface water are monitored for the presence of contaminants. The currently applied routine chemical analysis, albeit useful in many ways, gives no indication of the toxicity of pollutants and may miss unknown compounds that are harmful at

Bioluminescent Microbial Biosensors: Design, Construction, and Implementation
Edited by Gérald Thouand and Robert S. Marks

ISBN 978-981-4613-65-1 (Hardcover), 978-981-4613-66-8 (eBook)
www.panstanford.com

low levels. Biomonitors, such as algae and daphnia, do detect certain toxic compounds timely [1]. However, many harmful compounds that are very relevant from a human perspective, such as genotoxicants, cannot be measured by biomonitors. What is missing is an online system that sensitively and in real time detects compounds in water that are potentially harmful for human consumers.

A possible way to fill this gap is the use of genetically modified bacteria in which a promoter gene for a specific stress response is coupled to genes for luciferase (*lux*). By using different promoter genes, strains have been constructed for the detection of DNA, protein, and membrane damage; oxidative stress; several organic pollutants; and various heavy metals [2, 3]. The use of *luxCDABE* genes as a reporter gives strains the ability to emit luminescence without the addition of an external substrate [4]. Therefore these bacteria are very suitable for use in an online biosensor for the detection of toxic compounds in water.

Several bacterial biosensors have already been built that employ *lux* strains in combination with a photon detection device. These include miniature bioreactors with bacteria in suspension [5, 6], a sensor with freeze-dried bacteria [7], biochips [8], and self-contained sensors with bacteria immobilized onto optical fiber tips [9–11], later adapted as a flow-through sensor system [9].

It has been shown that it is possible to make a flow-through sensor based on luminescent bacteria that are immobilized in alginate gel on the tip of an optical fiber [9]. The latter was promising but not yet capable of long-term measurements while exhibiting limitations in sensitivity and sturdiness. For this reason, a new, improved prototype was built with some major adjustments. A dual-immobilization system was integrated, enabling the comparison of a larger glass slide conformation with that of the optical fiber. Immobilization on a larger surface has been done before in a disposable card, although in that system the light was brought to the photomultiplier through a fiber [12]. As only a small portion of the emitted light reaches the photomultiplier through the fiber [11], it was expected that the capture of the emitted light directly off a larger surface would increase the sensitivity.

However, since alginate gel does not chemically bind to a glass surface, it cannot be used to immobilize the bacteria on a glass slide. Therefore an inorganic sol-gel was used, which forms a porous, transparent silica layer. It has been shown that bacteria remain viable after immobilization in these gels [13–16]. Also, the response of genetically modified luminescent bacteria is not disrupted after immobilization [17]. To be able to compare the fiber and the plate, the sol-gel has been used for both setups.

The sensor also features control over temperature, flow speed, concentration of growth medium, and addition of positive controls. It was built with inert materials in order to prevent leaching of toxic compounds or memory effects from the sensor or tubing. To be able to ensure no living genetically modified bacteria are released into the environment, the efflux water was sterilized with either chlorine or ultraviolet (UV) light.

The new device was tested with the *Escherichia coli* strain DPD2794 [18], which responds to DNA damage and which is therefore one of the most relevant strains to use in such a device [3].

6.2 Description of the Strain and Growth Conditions

The bacterial strain used for testing the system was DPD2794 (kindly donated by Prof. S. Belkin through Prof. Marks [Ben-Gurion University, Beer-Sheva, Israel]). DPD2794 is a genetically modified *E. coli* strain that contains a plasmid with the *recA* promoter coupled to the *luxCDABE* genes. The *recA* gene is part of the bacterial SOS response, which utilizes DNA damage repair [18]. The *luxCDABE* genes are derived from *Aliivibrio fischeri* and code for the enzyme luciferase and the synthesis of its substrate. As a result of the coupling of these genes, DPD2794 generates light when DNA damage is induced [4]. The emitted light had a range of wavelengths from 400 nm to 750 nm, with a peak in intensity at 490–500 nm and a smaller second peak at 585–595 nm [19].

Stock cultures of bacteria were stored at –80°C in a 25% (v/v) glycerol medium. Work cultures were maintained at 4°C on Luria-Bertani (LB) agar plates, supplemented with 100 mg/L ampicillin. Before each experiment, bacteria were grown overnight in an incubator at 26°C (Friocell, MMM, Planegg, Germany). Bacteria were grown in an LB broth that contained 100 mg/L ampicillin.

The cultures were diluted to an OD_{600} of 0.2, as determined by a spectrophotometer (Jenway 6300 VIS, Staffordshire, England).

6.3 Immobilization of the Strain in a Sol-Gel

The sol-gel was prepared by mixing 4 mL tetramethyl orthosilicate (TMOS), 2.4 mL ultrapure water, and 0.68 mL 0.1 M HCl. This mixture was left for 24 h at 4°C to prepolymerize.

The glass slides were standard microscope slides (Waldemar Knittel, Braunschweig, Germany) purchased from Boom, Meppel. The optical fibers used were HUV-H pigtail fibers (Ceramoptec, Bonn, Germany). They had a core of pure, synthetic fused silica with a diameter of 1800 μm and a numerical aperture of 0.48 ± 0.2. The cladding was of a hard polymer with a diameter of 1900 μm. The tip on one end was polished; on the other end an SubMiniature version A (SMA) standard connector was attached.

The prepolymerized gel was first mixed with 0.05 M NaOH and then immediately mixed with the bacterial culture in a ratio of 1:1:4 sol-gel:NaOH:bacteria. The dilution caused a decrease in the OD_{600} of the mixture to 0.135 ± 0.01. One milliliter of the sol–bacteria mixture was immediately brought on the surface of the glass slide in a rectangular area of approximately 10 cm^2 and with a thickness around 1 mm. A drop with a height of approximately 3 mm was attached to the tip of the fiber by keeping the fiber upside down and letting the drop flow to the tip from a pipette (Fig. 6.1). Both slide and fiber were left for 5 min to let the gel polymerize.

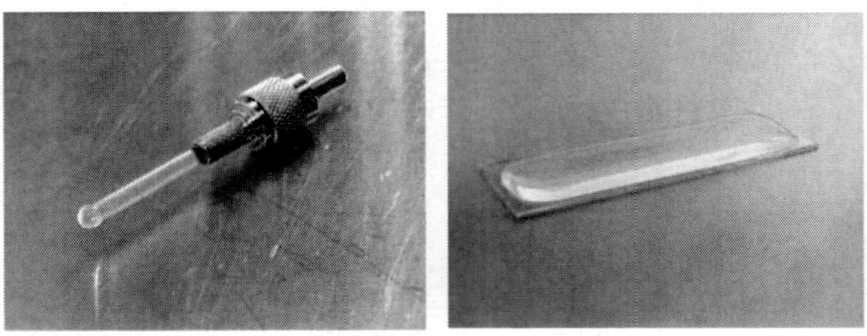

Figure 6.1 The bacteria are immobilized on the tip of an optical fiber and a microscope slide.

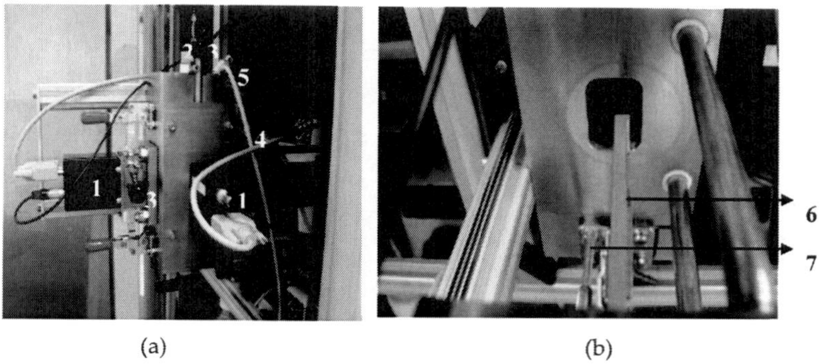

Figure 6.2 Photographs of the measurement chamber, as seen from the outside (a) and from the inside (b) of the pulled-out chamber side. 1: photomultiplier tubes; 2: pH and temperature sensor inlet; 3: light shutters; 4: water outlet; 5: air outlet; 6: plate; 7: fiber.

6.4 Construction of the Device

The glass plate and the optical fiber with the attached gel–bacteria mixture were fixed in a stainless steel measurement chamber (see Fig. 6.2).

This chamber was designed by 2M Sensors (Eindhoven, North Brabant, the Netherlands), and assemblage was done at KWR Watercycle Research Institute (Nieuwegein, Utrecht, the Netherlands). The outer dimensions of the body of the chamber were 70 × 85 × 155 mm. The inner volume of the chamber was reduced as much as possible, to a volume of 65 mL. This minimization was done to prevent a reduction or delay of the response by dilution of compounds in the chamber. In the enlargement in Fig. 6.3 the inner part of the measurement chamber is shown. The corners on the inside of the chamber were rounded both to reduce the volume and to ease the removal of biofilm. Tap water flowed from bottom to top through the chamber to prevent formation of air bubbles. The chamber was equipped with an air outlet at the top and a tap at the bottom to be able to drain the water. The top of the chamber also had an opening for a pH/temperature probe (Unitrode with Pt 1000, Metrohm, Schiedam, the Netherlands), which was used to measure the pH and temperature of the water directly in the chamber. Two

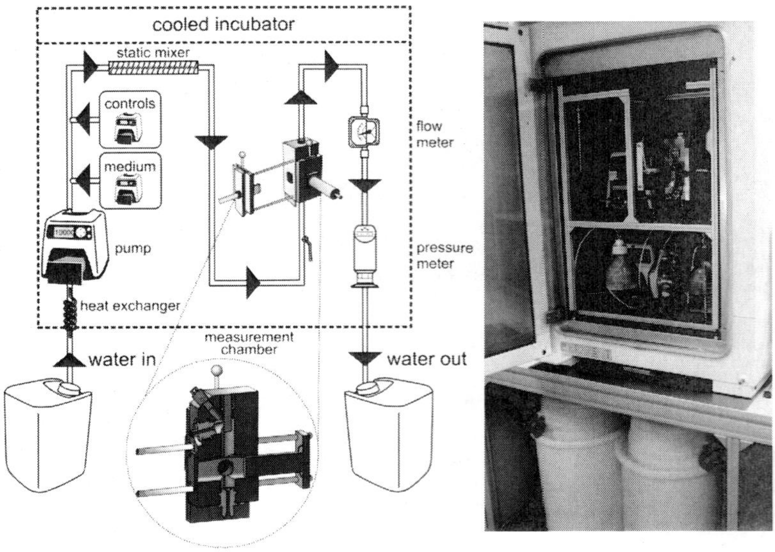

(a) (b)

Figure 6.3 An overview of the sensor. In (a) the outside of the measurement chamber is shown in the upper part, while the enlargement shows the inside of the chamber. A photograph of the sensor is shown in (b).

photomultiplier tubes (H-7467, Hamamatsu Photonics, Shimokanzo, Iwata, Japan) were attached to measure the light output of the bacteria on the plate and those on the fiber separately. Both were protected from overstimulation by manually operated light shutters. The chamber side to which the plate and the fiber were attached was easily removable to enable efficient changing of the bacteria (Fig. 6.2). Indirect lighting was shielded by the slide holder. In this way no light from the fiber could reach the slide and vice versa.

As can be seen in Fig. 6.3, the measurement chamber and all other parts of the sensor were placed in a cooled incubator (Friocell 222, MMM, Planegg, Munich, Germany). The only exceptions were the electronics and two 50 L water containers (Graf, Teningen, Germany). Before each experiment, one container was filled with tap water. The water that entered the sensor was first brought to 26°C with a heat exchanger inside the incubator.

The water was pumped through the system at 24.5 mL/min with a peristaltic pump (07523-80 L/S Digital Drive, Masterflex, Schiedam, the Netherlands) containing a preformed Teflon tube (77390-60, Masterflex, Schiedam, the Netherlands) with an inner diameter of 4 mm. A similar pump, but with Masterflex PharMed tubing, was used to add the LB medium at 0.5 mL/min to optimize the response of the bacteria. All tubing in the system, apart from the dosing of LB, was of either stainless steel or Teflon to prevent loss of compounds and memory effects. Masterflex PharMed tubing was used instead of Teflon for the addition of LB broth, because the pumphead necessary for Teflon tubing could not handle these small volumes. Possible slight absorption of the broth ingredients into the tubing was not so critical here, and experiments showed that this tubing did not leach any compounds that could affect the bacteria. The final concentration of the medium was either 7.5% or 2%, depending on the concentration of the medium in the bottle. The third pump was used to add compounds to the water and contained a Teflon tube (77390-50, Masterflex, Schiedam, the Netherlands) with an inner diameter of 2 mm. These compounds were used to verify the response of the bacteria and to characterize the system. Both the LB medium and the compounds were kept in 1 L glass bottles (Schott, Boom, Meppel, the Netherlands).

To ensure an even distribution of these additions, a static mixer (316 SS tube mixer, Cole-Parmer, Vernon Hills, Chicago, IL, USA) was installed before the measurement chamber. As an extra control measure the water passed through a flow meter (M-21, Tecfluid, Sant Just Desvem, Spain). Also a pressure meter (Ceraphant, Endress+Hauser, Reinach, Arlesheim, Switzerland) was installed after the measurement chamber as a safety measure. In case of an increase in pressure, the pumps were to stop automatically. In practice, however, it would have been better to install the pressure meter before the sensor chamber, because it was now behind most potential obstacles.

The water that left the sensor was collected in a waste barrel, where 1 L of bleach was added for disinfection (40 g chlorine per liter). In some experiments, the water went through an 18-watt UV-C water clarifier (AquaCristal Series II, JBL, Neuhofen, Rhein-Pfalz-Kreis, Rhineland-Palatinate, Germany), which was tested as an alternative method for disinfection. In these experiments, the UV unit was installed between the sensor and the waste barrel.

To clean the entire system, the chamber and the tubing in the sensor were flushed with household bleach diluted with tap water in a ratio of, respectively, 3:7 for approximately 15 min every day. Afterward, the system was thoroughly flushed with tap water for 1 h.

6.5 Electronics and Data Processing

All instruments that were part of the sensor were connected to a programmable automation controller (CompactRIO, National Instruments). The CompactRIO consists of a real-time controller, a chassis, and several I/O modules. Which I/O modules are necessary depends on the devices that have to be attached. In this case, the modules used were for RS232 (NI9870), Digital Input (NI9422), Analog Input (NI9203), Analog Output (NI9265), and Relay (NI9481). The CompactRIO and all cables were kept together in a metal box that was fixed on top of the incubator. The user input and data handling were performed on a Windows PC connected to the controller. The program that was used to control the sensor and gather the data was developed in Labview (version 2011, National Instruments). The luminescence was measured four times per minute, with an integration time of 10 ms. All results presented are averages over 5 min.

6.6 Influence of Nutrient Availability on the Response

The first priority after the construction of the new sensor was to establish a stable and reliable signal. An essential factor in achieving this was the concentration of nutrients available to the bacteria. In a previous study, it has already been shown that this strain of genetically modified *E. coli* did not respond without the addition of a medium to the water [9]. The optimum concentration found then was 7.5% LB medium, which was used as the starting point in the current study.

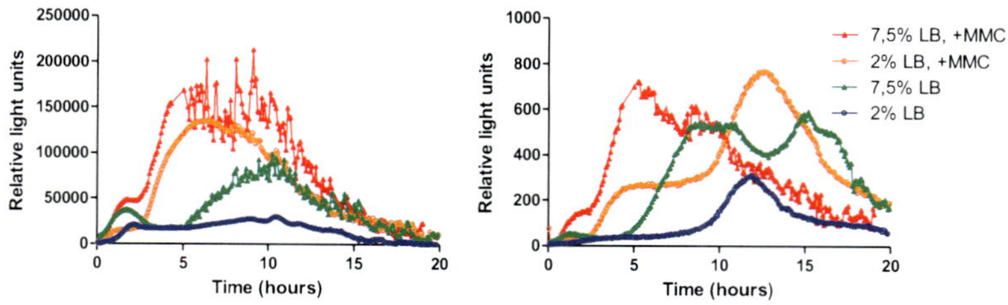

Figure 6.4 The effect of different nutrient concentrations on the response of the bacteria immobilized on a glass slide (left) or a fiber (right). In two curves in each graph, mitomycin C was added between $t = 1$ h and $t = 2$ h in a concentration of 0.1 mg/L.

The response with this medium concentration gave very strong fluctuations in the luminescence after about 5 h, both in the background and after exposure of the bacteria to 0.1 mg/L mitomycin C. This was accompanied by a strong increase in background luminescence, which might have been caused by growth of the bacteria. For this reason, the LB concentration was reduced to 2%. As can be seen in Fig. 6.4, a remarkable reduction of the fluctuations was observed at this concentration without loss of the response on the plate. On the fiber, there was some reduction in response at the beginning of the curve, but the response was still clearly distinguishable. Also, the contrast between the background and the response levels increased as a result of a decrease in background luminescence. Only on the fiber, a peak remained after about 10 h. The cause of this peak remains unclear, especially since it did not occur on the plate. The observed lag time, the time between the start of the dosing and the start of the response, remained the same as with 7.5% LB, namely about 1 h. An additional advantage of a lower medium concentration was that it delayed the manifestation of biofouling in the system. In all the following experiments, 2% LB was therefore used as the nutrient source.

6.7 The Effect of Biofouling

In all previously shown experiments, the sensor had priorly been cleaned with diluted bleach to remove biofouling from the system.

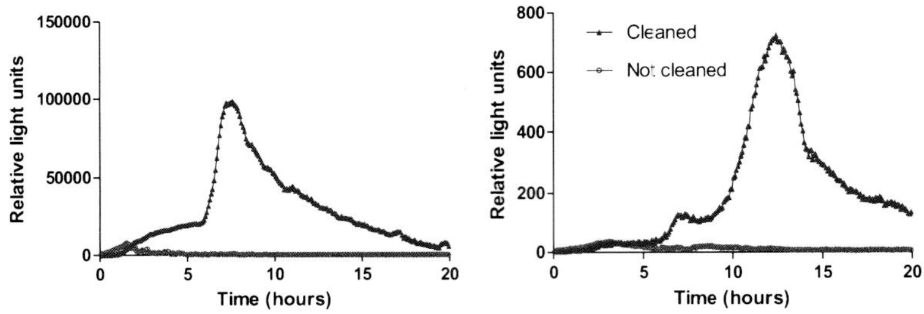

Figure 6.5 The difference in response when using the sensor directly after a previous experiment and when the system has been cleaned in between. The bacteria were exposed to 1 mg/L nalidixic acid between $t = 5$ h and $t = 6$ h. The response of the glass slide is depicted on the left and the response of the fiber on the right.

However, some experiments were also performed without prior cleaning, continuing directly with the new measurement after the previous measurement. In these cases, the signals generated on both the fiber and the glass slide were drastically altered in comparison with the signals gained from a clean system.

In Fig. 6.5, an example is given of an experiment where nalidixic acid was administered as a positive control between 5 and 6 h. In one case the system had been cleaned, and in the second case the experiment was performed directly after a 24 h experiment, without any in-between cleaning or flushing. In the first case, there was a normal background, followed by a sharp increase in the signal in response to the positive control. Without cleaning, the bacteria on the plate gave a relatively low and short peak, after which the response decreased. Also on the fiber, the response was lower and without the characteristic peak between 600 and 800 min when the system had not been cleaned first. Similar experiments without cleaning showed the same results, with a much lower or even absent response (results not shown). Also very typical in these instances was the occurrence of strong fluctuations in luminescence, especially on the glass slide.

The consequence of this drastic effect of fouling was that the sensor had to be cleaned daily between the experiments. Some measures might be taken in future sensors to reduce this problem. An important improvement would be to add the medium as close to the bacteria as possible, since most fouling occurred from the point where the medium was added to the water. One can also think of using filters to remove the bacteria from the water, especially once a sensor is used with surface water.

6.8 Determination of Sensitivity

An important aspect of the sensor is the achieved sensitivity, as the device can only be useful in the field if it can detect the relevant contamination levels. To test the sensitivity of the sensor device, the bacteria were exposed to several concentrations of mitomycin C, the most potent compound found so far for this strain, between 1 and 2 h from the start of the experiment. The signal started to differ from the background about 1 h after the start of the exposure, thus about 2 h from the start of the experiment. To determine the limit of detection (LOD), the responses of the exposed bacteria were compared with the average of three background curves. For the concentrations of 0.01 mg/L and 0.001 mg/L, the average of two experiments was plotted.

The choice was made not to use a response ratio (RR) (response divided by the background) of two as the LOD, as is usually done. The reason was that this can render an analysis unnecessarily insensitive in the case of high backgrounds with relatively small variation or more sensitive than realistic in cases with high variation. In this respect, Takanashi and Uranako (1998) and Kim and Margolin (1999) considered the application of the conventional twofold rule to the data of Ames tests as too conservative (i.e., few samples are evaluated as genotoxic) [20, 21].

A common evaluation method in analytical chemistry, which is suitable for normally distributed data, is to define a response as significantly different from the background when it exceeds the average response of the background plus three times the standard deviation of the background response. This method is also compulsory to comply with the ISO standard for the specification and performance tests of new online sensors and analyzing equipment for water [22]. For normally distributed data, there is then 99% certainty that the sample response is significantly

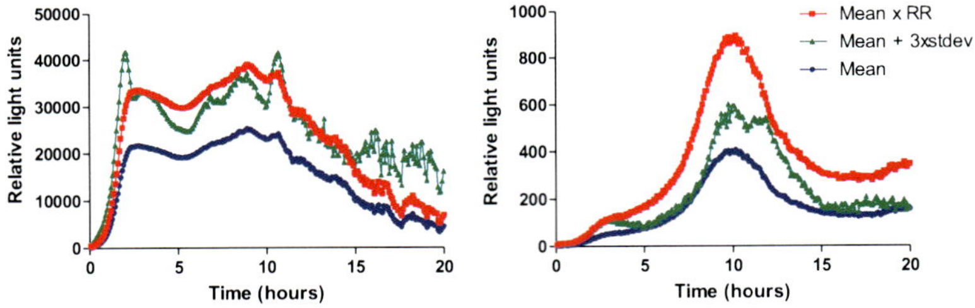

Figure 6.6 An overview of the average background of two experiments, the background plus three times the standard deviation, and the average times the highest response ratio (RR) of the previous two curves between 2.5 and 10 h. This ratio was 1.55 for the glass slide and 2.19 for the fiber. On the left is the response of the bacteria on the glass slide and on the right the response of the fiber.

different from the background. We have therefore chosen to apply an adapted version of this latter method.

To correct for the variation between experiments, a curve was made of the average background of three experiments, plus three times the standard deviation (see Fig. 6.6). This curve was then divided by the average background to obtain an RR. The highest RR between 2.5 and 10 h was determined as a significance limit independent of the point in time; this was 1.55 for the plate and 2.19 for the fiber. The curve of the average background response times this highest RR was the significance limit (Mean × RR in Figs. 6.6 and 6.7). The LOD was defined as the lowest concentration that gave a response that was above this significance limit (i.e., 1.55 or 2.19 times the average background) anywhere in the time period between 2.5 and 10 h of monitoring time, as visualized in Fig. 6.7. Using this definition, the LOD for detecting mitomycin C was estimated to be 0.01 mg/L for the plate setup and 0.1 mg/L for the fiber configuration of the sensor. The higher LOD of the fiber was mainly caused by the variation in the background peak of this setup. Especially the variation in the time at which this peak occurred reduced the sensitivity of the fiber. The variation on the plate was relatively low compared to the background, resulting in a higher sensitivity.

The newly added plate setup clearly is an improvement to the design of an online bacterial toxicity sensor using these modified

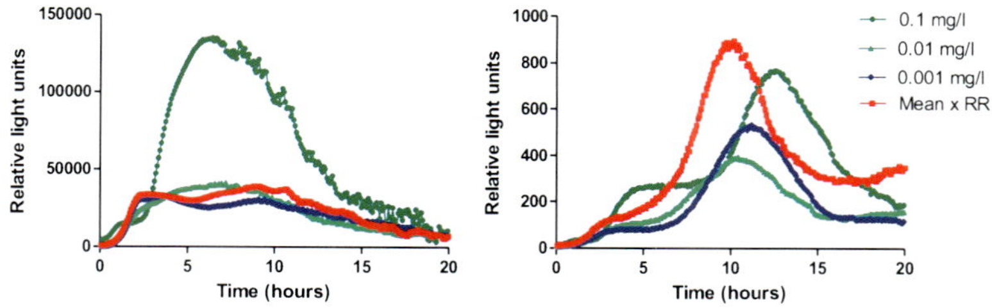

Figure 6.7 The response to different concentrations of mitomycin C, which was administered between 1 and 2 h, and the curve of the average background response times the response ratio of, respectively, 1.55 for the slide and 2.19 for the fiber. On the left is the response of the bacteria on the glass slide and on the right the response of the fiber.

bacteria. The detection of 10 μg/L of a genotoxic substance by the plate setup lies in the same range as the sensitivity of other biomonitors. The sensitivity of the daphnia monitor lies between 1 and 10 μg/L for the insecticides it is most sensitive for. The algae monitor has an LOD for herbicides of 0.5–2 μg/L, depending on the compound [23]. The current sensitivity of the sensor is sufficient to detect heavy plumes of contamination in surface water. Unfortunately, it is not yet sensitive enough to detect compounds at the target values for drinking water, which are 10 ng/L for genotoxic compounds and 0.1 μg/L for other compounds [24]. Additional improvement is necessary, possibly achievable through reduction of the variation in the response by further standardization or through online sample preconcentration.

It is also noteworthy that the response to mitomycin C endured long after the inducing compound was released from the system, especially after exposure to a concentration of 0.1 mg/L. High-performance liquid chromatography (HPLC) analysis of samples from the sensor showed that no mitomycin C was detectable 30 min after stopping the input of the toxicant. This indicates that the signal was caused by a prolonged biochemical reaction of the DNA damage repair system. The implication for field monitoring is that once a positive signal has been induced, either the bacteria have to be replaced or a long lag period must be taken until the bacteria return to their normal state.

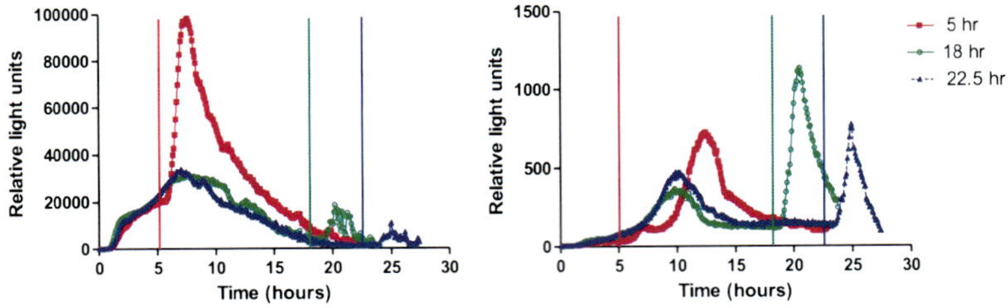

Figure 6.8 The response of the bacteria on the glass slide (left) and the fiber (right) to 1 mg/L nalidixic acid administered for 1 h after 18 or 22.5 h.

6.9 Delayed Exposure

The ability of the bacteria to respond at a later time point was tested by dosing 1 mg/L nalidixic acid after 5, 18, or 22.5 h in separate experiments. A clearly distinguishable response was observed both on the fiber and on the glass slide at all time points (Fig. 6.8). Very striking was the response at 5 h on the plate, which was much higher than the response at later time points. This was very different from the fiber, which gave a low response at 5 h, but showed almost no decrease over time. It may be possible that the same processes that give the background peak between 5 and 15 h on the fiber also result in a lower response during this time period.

Another observation regarding the glass slides was the occurrence of typical fluctuations in the response. These were already present before the addition of the positive controls but became more pronounced during the response peak. It has been thought that this might be caused by the visually observed growth of other microorganisms on the gel during the experiment, which might hinder the luminescent *E. coli*. What is the cause of these differences between both setups, which entail the same bacteria and sol-gel, remains a question still to be answered.

6.10 Comparison of Immobilization Methods

The bacteria were immobilized on both a glass slide and an optical fiber, which were used simultaneously. Fibers have been used in previous studies. Thus they were already known as a viable platform for immobilization of bacteria and guidance of the produced light to a detector [9, 25]. In this study, fibers were used with a larger diameter than previously used, namely 1800 μm instead of 400 μm. The motivation behind this choice was that with a larger surface of the tip, more bacteria are directly in front of the fiber. Therefore a larger portion of the emitted light enters the fiber. It was also for this reason that glass slides were tested as an immobilization platform, as they have a larger surface area than fibers. The idea was that by being able to gather more light, the resolution of the signal as well as the sensitivity would increase.

As can be seen in all figures, the signal from the plate is often more than a factor of 100 higher than that of the fiber. This difference coincides well with the difference in the surface area from which light was captured, which was 5 mm^2 for the fiber and approximately 1000 mm^2 for the plate.

More unexpected was the difference in shape of the background response from the bacteria. The bacteria on the plate always started with a steep rise in response, followed by a slow decrease after about 10 h. The response of the bacteria on the fiber increased more slowly and showed a characteristic peak between 600 and 800 min from the start of the experiment. This pattern remained when a mutagenic compound was added before the peak, although an increase in light production was observed. In comparison, the response of the plate was sharper and probably easier to observe when the exact time of the dosing was not known.

More importantly, the LOD of the plate was lower, namely 0.01 mg/L versus 0.1 mg/L for the fiber. This difference was caused by the high variation in the height and time of the background peak of the fiber. It should be noted, though, that the response of the fiber improved, when the exposure took place later, after the response of the fiber had flattened out. As can be seen in Fig. 6.8, exposure after 18 h gave a much sharper response in the fiber than with the plate at that time or with the fiber at an earlier time. Nevertheless, if one setup would have to be chosen for use in the field, depending on these results, the plate is the better choice due to its higher sensitivity and predictability.

It is still unknown which underlying mechanisms determine the size and shape of background luminescence. A theory is that the luminescence is the result of the activation of DNA repair during normal growth of the bacteria. In that case, the decline of the signal in time could be caused by limitations of space in the sol-gel and competition both among the *E. coli* themselves and with other microorganisms. However, this does not explain why the signal is different depending on whether bacteria are immobilized on a fiber or on a plate. A factor that might have to do with this is that the light from the fiber is measured from the bottom of the gel, while the light from the slide is measured from the surface. In the case of the fiber, mass diffusion of compounds and oxygen through the gel could be a limitation, as already shown for agarose-immobilized bacteria [26]. An effect can also be expected from the difference in absorption and scattering of the light when it travels through either the gel on the fiber or the water column between the plate and the photomultiplier.

It would be interesting to see whether the relatively low response of the plate in longer experiments is related to fouling with foreign microorganisms. If this is the case, it would be expected that the response improves when the natural bacteria are removed by, for example, filtration. A related question that needs to be answered is what the maximal life span is of one batch of bacteria and whether the immobilization matrix has an influence on this.

6.11 Evaluation of the UV Disinfection Unit

In several countries, the use of genetically modified bacteria is bound to strict regulations. This means that all contaminated material, including the wastewater from the sensor, has to be disinfected. The use of chlorine bleach has been established as a valid method for the disinfection of water contaminated with genetically modified *E. coli* [9]. However, this method has the disadvantages that it cannot be used online and that the use and disposal of large quantities of chlorine is undesirable. Therefore, UV-C irradiation was tested as an alternative method of disinfection.

A water clarifier originally meant for aquaria was tested, in which the water flowed past an 18-watt UV-C lamp. Three obstacles around the lamp prevented the water from taking the shortest route, thus preventing some parts of the water from being irradiated. The UV unit was installed directly after the sensor and before the waste barrel and tested during several experiments with the sensor.

The effectiveness of the disinfection was determined by collecting 0.1 mL samples before and after the UV unit, as well as 100 mL after the UV lamp, and plating these on LB agar plates. The collected plate counts from three separate experiments showed an average number of 94 colonies per plate before the UV unit, which corresponds with 9.4×10^4 bacteria per milliliter. No colonies were found on any of the plates with the samples taken after the UV unit, including that of the filtered 100 mL sample. This indicated that the UV unit was very effective in inactivating the bacteria (>log 6 removal), at least under laboratory conditions. It should be mentioned, though, that the effectiveness of UV can decline when the UV lamp ages or when the water is more turbid. Therefore, although UV was a very useful method for disinfection, regular monitoring remained necessary.

6.12 Conclusions

A new sensor was developed and tested for the online measurement of genotoxic compounds in water with luminescent bacteria. In this sensor, bacteria for the detection of DNA damage were immobilized on an optical fiber and a glass slide. The light signal from the bacteria on the glass slide was much stronger than from the fiber, as a result of the larger surface area. Moreover, the variation between experiments was higher with the fiber, which reduced the sensitivity. The detection limit was 0.1 mg/L for the fiber and 0.01 mg/L for the glass slide. To increase the sensitivity in the future, it is important to reduce the variation as far as possible. A response could still be measured after 22 h. However, the height of the response, as well as the background luminescence, decreased over time. This was probably caused by interference of other microorganisms. High numbers of foreign bacteria also disabled the luminescent bacteria when the sensor was used for more than 1 day without cleaning. If this problem is solved, this sensor could be a useful addition to the current biomonitors for the screening of surface water.

Acknowledgments

This work was supported by the Joint Research Program of the Dutch water utilities (BTO) and was performed at KWR within the TTIW-cooperation framework of Wetsus, Centre of Excellence for Sustainable Water Technology (www.wetsus.nl). Wetsus is funded by the Dutch Ministry of Economic Affairs. The authors would like to thank the participants of the sensoring research theme for the fruitful discussions and their financial support.

References

1. De Hoogh, C. J., Wagenvoort, A. J., Jonker, F., Van Leerdam, J. A., Hogenboom, A. C. (2006). HPLC-DAD and Q-TOF MS techniques identify cause of Daphnia biomonitor alarms in the River Meuse. *Environ. Sci. Technol.*, **40**, 2678–2685.
2. van der Meer, J. R., Belkin, S. (2010). Where microbiology meets microengineering: design and applications of reporter bacteria. *Nat. Rev. Microbiol.*, **8**, 511–522.
3. Woutersen, M., Belkin, S., Brouwer, B., Van Wezel, A. P., Heringa, M. B. (2011). Are luminescent bacteria suitable for online detection and monitoring of toxic compounds in drinking water and its sources? *Anal. Bioanal. Chem.*, **400**, 915–929.
4. Billard, P., DuBow, M. S. (1998). Bioluminescence-based assays for detection and characterization of bacteria and chemicals in clinical laboratories. *Clin. Biochem.*, **31**, 1–14.
5. Lee, J. H., Gu, M. B. (2005). An integrated mini biosensor system for continuous water toxicity monitoring. *Biosens. Bioelectron.*, **20**, 1744–1749.
6. Thouand, G., Horry, H., Durand, M. J., Picart, P., Bendriaa, L., Daniel, P., DuBow, M. S. (2003). Development of a biosensor for on-line detection of tributyltin with a recombinant bioluminescent Escherichia coli strain. *Appl. Microbiol. Biotechnol.*, **62**, 218–225.

7. Choi, S. H., Gu, M. B. (2002). A portable toxicity biosensor using freeze-dried recombinant bioluminescent bacteria. *Biosens. Bioelectron.,* **17**, 433–440.
8. Yoo, S. K., Lee, J. H., Yun, S. S., Gu, M. B. (2007). Fabrication of a bio-MEMS based cell-chip for toxicity monitoring. *Biosens. Bioelectron.,* **22**, 1586–1592.
9. Eltzov, E., Marks, R. S., Voost, S., Wullings, B. A., Heringa, M. B. (2009). Flow-through real time bacterial biosensor for toxic compounds in water. *Sens. Actuators B,* **142**, 11–18.
10. Hakkila, K., Green, T., Leskinen, P., Ivask, A., Marks, R., Virta, M. (2004). Detection of bioavailable heavy metals in EILATox-Oregon samples using whole-cell luminescent bacterial sensors in suspension of immobilized onto fibre-optic tips. *J. Appl. Toxicol.,* **24**, 333–342.
11. Polyak, B., Bassis, E., Novodvorets, A., Belkin, S., Marks, R. S. (2001). Bioluminescent whole cell optical fiber sensor to genotoxicants: system optimization. *Sens. Actuators B,* **74**, 18–26.
12. Horry, H., Charrier, T., Durand, M. J., Vrignaud, B., Picart, P., Daniel, P., Thouand, G. (2007). Technological conception of an optical biosensor with a disposable card for use with bioluminescent bacteria. *Sens. Actuators B,* **122**, 527–534.
13. Fennouh, S., Guyon, S., Livage, J., Roux, C. (2000). Sol-gel entrapment of Escherichia coli. *J. Sol-Gel Sci. Technol.,* **19**, 647–649.
14. Kuncova, G., Podrazky, O., Ripp, S., Trogl, J., Sayler, G. S., Demnerova, K., Vankova, R. (2004). Monitoring of the viability of cells immobilized by sol-gel process. *J. Sol-Gel Sci. Technol.,* **31**, 335–342.
15. Desimone, M. F., De Marzi, M. C., Copello, G. J., Fernandez, M. M., Pieckenstain, F. L., Malchiodi, E. L., Diaz, L. E. (2006). Production of recombinant proteins by sol-gel immobilized Escherichia coli. *Enzyme Microb. Technol.,* **40**, 168–171.
16. Alvarez, G. S., Foglia, M. L., Copello, G. J., Desimone, M. F., Diaz, L. E. (2009). Effect of various parameters on viability and growth of bacteria immobilized in sol-gel-derived silica matrices. *Appl. Microbiol. Biotechnol.,* **82**, 639–646.
17. Premkumar, J. R., Rosen, R., Belkin, S., Lev, O. (2002). Sol-gel luminescence biosensors: Encapsulation of recombinant E-coli reporters in thick silicate films. *Anal. Chim. Acta,* **462**, 11–23.
18. Vollmer, A. C., Belkin, S., Smulski, D. R., Van Dyk, T. K., LaRossa, R. A. (1997). Detection of DNA damage by use of Escherichia coli carrying recA'-lux, uvrA'-lux, or alkA'-lux reporter plasmids. *Appl. Environ. Microbiol.,* **63**, 2566–2571.

19. Thouand, G., Daniel, P., Horry, H., Picart, P., Durand, M. J., Killham, K., Knox, O. G. G., Dubow, M. S., Rousseau, M. (2003). Comparison of the spectral emission of lux recombinant and bioluminescent marine bacteria. *Luminescence,* **18**, 145–155.

20. Kim, B. S., Margolin, B. H. (1999). Statistical methods for the Ames Salmonella assay: a review. *Mutat. Res.,* **436**, 113–122.

21. Takanashi, H., Urano, K. (1998). Statistical procedures for estimating the detection limit and determination limit of the Ames Salmonella mutagenicity assay. *Sci. Total Environ.,* **221**, 31–42.

22. ISO, I. O. F. S. (2003). *Water Quality: On-line Sensors/Analysing Equipment for Water; Specifications and Performance Tests* (ISO 15839).

23. Green, U., Kremer, J. H., Zillmer, M., Moldaenke, C. (2003). Detection of chemical threat agents in drinking water by an early warning real-time biomonitor. *Environ. Toxicol.,* **18**, 368–374.

24. van Leeuwen, C. J., Mons, M., Heringa, M. B., Puijker, L. M., van der Kooij, D., Stoks, P. G. M., van der Hoek, J.-P. (2013). Use of the threshold of toxicological concern (TTC) approach for deriving target values for drinking water contaminants. *Water Res.,* **47**(4), 1666–1678.

25. Polyak, B., Bassis, E., Novodvorets, A., Belkin, S., Marks, R. S. (2000). Optical fiber bioluminescent whole-cell microbial biosensors to genotoxicants. *Water Sci. Technol.,* **42**, 305–311.

26. Affi, M., Solliec, C., Legentillomme, P., Comiti, J., Legrand, J., Thouand, G. (2009). Numerical design of a card and related physicochemical phenomena occurring inside agarose-immobilized bacteria: a valuable tool for increasing our knowledge of biosensors. *Sens. Actuators B,* **138**, 310–317.

Chapter 7

Development and Characterization of a Living-Cell Bioluminescent Bioreporter Integrated Circuit

James T. Fleming,[a] Syed Islam,[b] Nora Dianne Bull,[c] Michael Simpson,[a,d,e] and Gary Sayler[a]

[a] *Center for Environmental Biotechnology, University of Tennessee, Knoxville, TN 37996, USA*
[b] *Department of Electrical and Computer Engineering, University of Tennessee, Knoxville, TN 37996, USA*
[c] *Oak Ridge National Laboratory, Microelectronics Systems Research Group, Oak Ridge, TN 37831, USA*
[d] *Oak Ridge National Laboratory, Center for Nanophase Materials Sciences, Oak Ridge, TN 37831, USA*
[e] *Department of Materials Science and Engineering, University of Tennessee, Knoxville, TN 37996, USA*
jtf@utk.ed

7.1 Introduction

Whole-cell biosensors incorporate living cells as integral components of a sensing device. The biological elements may be natural prokaryotic or eukaryotic cells or genetically engineered variants

Bioluminescent Microbial Biosensors: Design, Construction, and Implementation
Edited by Gérald Thouand and Robert S. Marks

ISBN 978-981-4613-65-1 (Hardcover), 978-981-4613-66-8 (eBook)
www.panstanford.com

thereof, isolated proteins, including antibodies, or other isolated cellular components. These biological elements are then integrated in some fashion to an electronic device, which then (measures or) transduces the biological signal into a form that can be directly analyzed, stored, or transmitted (for further analysis). A variety of transducing methods have been developed, including colorimetric, fluorescent, electrochemical, and bioluminescent approaches [1, 2]. The use of bioluminescent reporters has been widely adopted for chemical sensing because of their sensitivity, specificity, range, relative simplicity, and suitability for real-time analysis [3]. Here prokaryotic or eukaryotic cells are engineered to emit light in response to an environmental stimulus. The genetic engineering usually involves inserting the promoter of a gene that is naturally upregulated in response to a specific stimulus, in front of luciferase structural genes. Thus under the conditions that upregulate a specific gene, gene expression of the luciferase genes is induced and light is emitted. As it is conventionally practiced the emitted light is generally detected by a benchtop photodetector instrument that typically uses a photomultiplier tube (PMT). With the initial goal of creating a mobile device, we have miniaturized the light detection apparatus using complementary metal-oxide semiconductor (CMOS) technology to make a bioluminescent bioreporter integrated circuit (BBIC).

7.2 Development of the Integrated Circuit Component

The integrated circuit (IC) contains the individual circuit elements that (1) detect the optical (light) signal generated by the bioreporter and (2) process and digitize the signal. A block diagram illustrating the general device components of the first-generation BBIC is shown in Fig. 7.1.

The photodetector consists of an array of photodiodes implemented in a standard CMOS process. The n-well/p-substrate CMOS photodiode was chosen because, compared to other photodetector designs, its quantum efficiency is better [4]. The signal-processing system converts the photodetector current into a digital signal by way of a current-to-frequency converter (CFC) circuit. The main

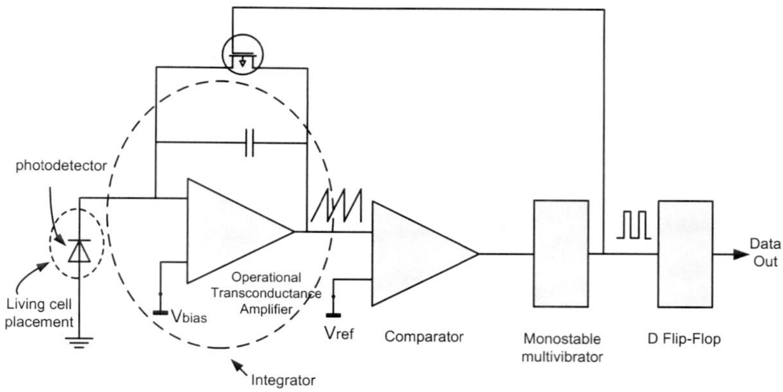

Figure 7.1 General schematic of the first-generation BBIC. The main system components are the photodetector and the current-to-frequency converter (CFC). The CFC is composed of an integrator, a comparator, a one-shot, and a gated counter. The current output from the integrator rises until the V_{ref} of the comparator is reached, which, in turn, triggers a pulse from the multivibrator. This pulse resets the integrator and generates a frequency signal that is proportional to the photodetector current. Living cells are positioned over the photodetector. Adapted from Ref. [4] with permission.

CFC components include an integrator, a comparator, a monostable multivibrator, and a D flip-flip (Fig. 7.1). The integrator consists of an operational transconductance amplifier (OTA) in parallel with a switched capacitor. The integrator collects the photodetector current until the voltage reaches a threshold voltage, V_{Th}, which is equal to the comparator threshold, V_{TH}. At this point the output of the comparator changes from low to high, which, in turn, triggers an output pulse from the monostable multivibrator. The pulse causes the T-flip-flop-gated counter to change states, and the frequency of the pulse is obtained by way of an exterior counter/timer. The output from the multivibrator also resets the integrator by closing a metal-oxide semiconductor field-effect transistor (MOSFET) switch across the integration capacitor, discharging it and resetting the output of the integrator to the baseline level (V_{Bias}). This sequential process produces a series of digital pulses at a frequency that is proportional to the photodetector current.

The evolution of the BBIC design has been driven, to a large extent, by the desire to achieve a high degree of sensitivity. In this respect the minimum detectable signal (MDS) is an important parameter as it determines the low-end sensitivity of the system [5]. It is a function of the quantum efficiency, the leakage current, the noise of the photodetector, and the noise and filtering characteristics of the signal-processing circuit. The MDS may be expressed as

$$\text{MDS} = \frac{1}{q\eta}\sqrt{\frac{4qA_{\text{PD}}I_{\text{s}}}{T_{\text{int}}}} \tag{7.1}$$

where q is the electronic charge, η is the quantum efficiency, A_{PD} is the area of the photodiode, T_{int} is the integration time, and I_{s} is the photodiode saturation current. This equation is valid only if the bias on the photodiode is zero and the noise of the signal-processing circuitry is negligible compared with the photodiode noise. Thus to increase the detection limit (e.g., low MDS), $\eta\eta$ and T_{int} should be increased and I_{s} and A_{PD} should be minimized. I_{S} can be minimized by optimizing the protective coating on the chip. Practical considerations limit the T_{int}, and a small A_{PD} puts a prohibitive limitation on light collection.

From a practical standpoint photodetector currents are obtained by measuring the frequency of the output pulses as

$$f_{\text{out}} = \frac{I_{\text{P}}}{2C_{\text{int}}\Delta V} \tag{7.2}$$

where f_{out} is the frequency of the output pulses, I_{P} is the photodetector current, C_{int} is the integration capacitance, and ΔV is the threshold voltage of the integrator. The photocurrent can be calculated as

$$I_{\text{P}} = \frac{2 \cdot N \cdot C_{\text{int}} \cdot \Delta V}{\text{T}} \tag{7.3}$$

where N is the number of pulses counted over a particular time. The shapes of the waveforms at various points in the system are also shown in Fig. 7.1.

7.2.1 *The First-Generation BBIC*

A block diagram of the first-generation BBIC chip is shown in Fig. 7.1. In this design a single MOSFET switch was used to reset the integrator, as described in Section 7.2. It used an optical application–specific integrated circuit (OASIC) mounted in a 40-pin ceramic

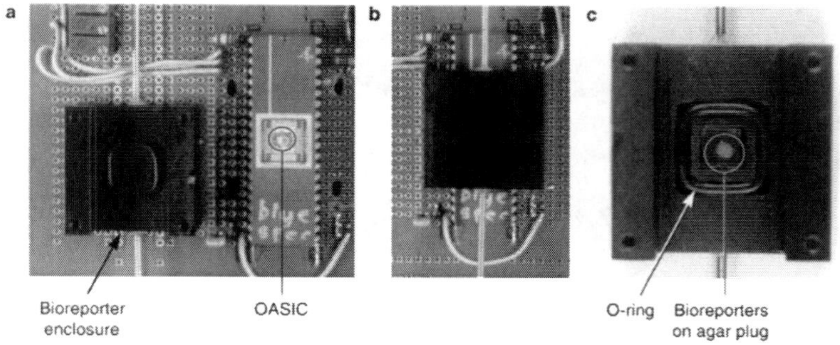

Figure 7.2 The first-generation BBIC. (a) The optical application–specific integrated circuit (OASIC) mounted in a 40-pin ceramic chip carrier and the enclosure for the cells. (b) The enclosure mounted on the chip. (c) The enclosure with bioreporters on an agar plug. The O-ring is used to make a light-tight seal between the ceramic chip carrier and the bioreporter enclosure. Reproduced from Ref. [5] with permission.

chip carrier over which was placed an enclosure supporting the bioreporter in an agar plug, as shown in Fig. 7.2 [5]. The OASIC measured 2.2 × 2.2 mm and was fabricated in a standard 1.2 μm n-well CMOS IC process.

The OASIC was coated with a protective silicon nitride film deposited using the molecular jet chemical vapor deposition process. The signal-conditioning circuit produced an output pulse with a frequency directly proportional to the photodiode current. A signal proportional to the sum of the leakage current and the photocurrent was generated, integrating pulses for a specific time. Compared to conventional electrometer circuits the signal-conditioning circuit provided lower noise, faster recovery from overloads, and a larger dynamic range. With an integration time of 1 min and no cells present the leakage current produced a signal of approximately six counts per minute (cpm). The standard deviation decreased with the square of root of the integration time. The toluene-sensitive bioreporter *Pseudomonas putida* TVA8, known to have a linear light response to toluene to saturation at 10 parts per million (ppm), was placed on the OASIC chip and exposed to toluene vapor estimated to represent a concentration of ~1 ppm. Under these conditions a signal of 12 cpm was generated (6 above background), with a standard deviation (σ) of 0.22 cpm. Assuming

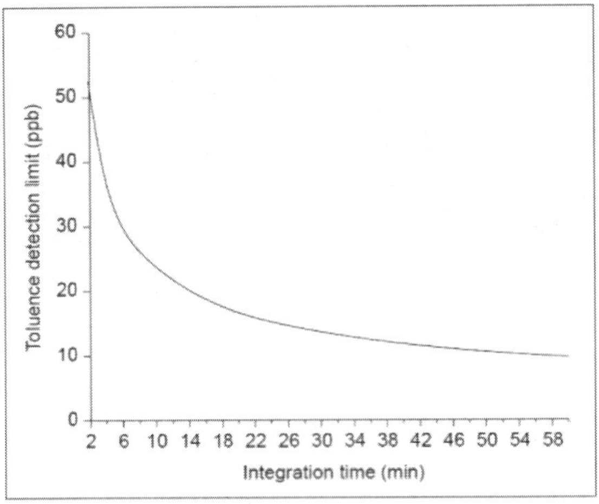

Figure 7.3 MDC of toluene as a function of integration time for the prototype BBIC employing *Pseudomonas putida* TVA8 bioreporter. Reproduced from [5] with permission.

that an MDS is 2σ above background (6.44 cpm) the minimum detectable concentration (MDC) of toluene in parts per billion (ppb) is given by

$$\text{MDC} = \frac{(10,000)\,6.44}{\sqrt{T}_{\text{int}} S_{\text{sat}}} \tag{7.4}$$

where S_{sat} is the signal in cpm at a concentration of 10 ppm and T_{int} is the integration time in minutes. The minimum detectable toluene concentration for this BBIC as a function of integration time approaches 10 ppb, as shown in Fig. 7.3 [5].

7.2.2 *The Second-Generation BBIC*

The second-generation BBIC chip improved on the first version by optimizing the microluminometer for (1) detection of low-level bioluminescence and (2) minimization of photodetector temperature-related artifacts. This was achieved by carefully regulating the subfemtoampere signal and the leakage currents in the front-end processing circuitry [4]. Here the photodiode was operated with a reverse bias of only a few millivolts. The amount of leakage current

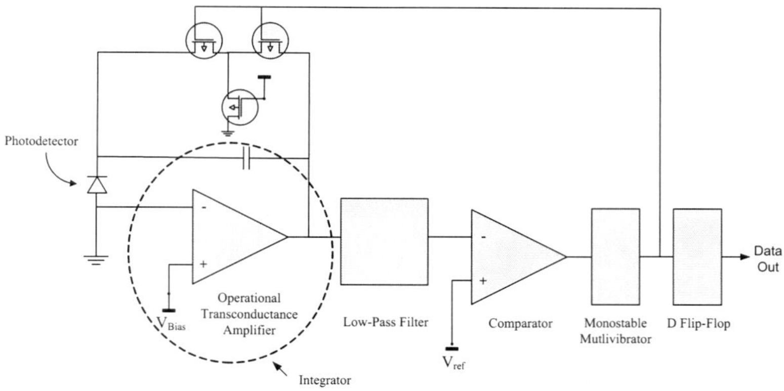

Figure 7.4 Schematic of the second-generation BBIC. Operating the photodetector at a low bias required the addition of two extra MOSFET switches. Adapted from Ref. [4] with permission.

can be estimated by the ideal diode equation

$$I_d = I_s\left(e^{V_A/V_T} - 1\right) \tag{7.5}$$

where I_d is the forward diode current, I_S the reverse saturation current, V_A the applied bias voltage, and V_T the thermal voltage. The leakage current is directly proportional to the junction area and approximately doubles for each 5°C increase in temperature. As the reverse current is indistinguishable from the signal current, operating the photodiode with a very low reverse bias will minimize the temperature effect on the reverse current. Operating the reverse bias at such a low level requires resetting the CFC. At a low reverse bias on the photodetector, the subthreshold current of the N-type metal-oxide semiconductor (NMOS) switch exceeded the leakage current of the photodiode.

Thus with low signals the current flow into the integration capacitor was in the wrong direction, resulting in a limited MDS. This problem was addressed by replacing the single NMOS transistor switch with two switches in series and a third connected between the first two (T-switch) (Fig. 7.4) [4].

In this configuration the leakage current flows to ground and very little through the device. A low-pass filter was added between the integrator and the comparator to remove high-frequency noise from the integrator and the photodetector. With this design, at 2 mV (the lowest reverse bias used) an MDS of 0.15 fA was measured for an integration time of 1510 s. If a single fully induced *Pseudomonas fluorescens* 5RL cell produces approximately 3×10^{-20}A, then the MDS can be expressed as ~5000 fully induced cells [4].

7.2.3 *The Third-Generation BBIC*

For this, the third and the most current, BBIC version, a further series of enhancements were made to improve robustness and system performance [6, 7]. The BBIC circuit was redesigned in a 0.35 μm CMOS process to (1) reduce the power consumption compared to the first and the second iterations and (2) accommodate an increased number of photodiodes for a given area, thus increasing the sensitivity. The total power consumption is now 3 mW from a 3.3 V power supply compared to the earlier versions, which used 15 mW from a 5 V power supply. The integration reset technique was further modified to ensure that the OTA operates in high-gain mode at the start of the integration process. Switches S1 and S2 set the integrator input and output to ground and 0.5 V, respectively, ensuring that the OTA is in the high-gain region of operation. The OTA was improved for an open-loop gain of 92 dB compared to an open-loop gain of 75 dB in the second version. The generation of bias voltages was achieved for all the system blocks from a single bandgap reference circuit. This cell takes a bandgap reference

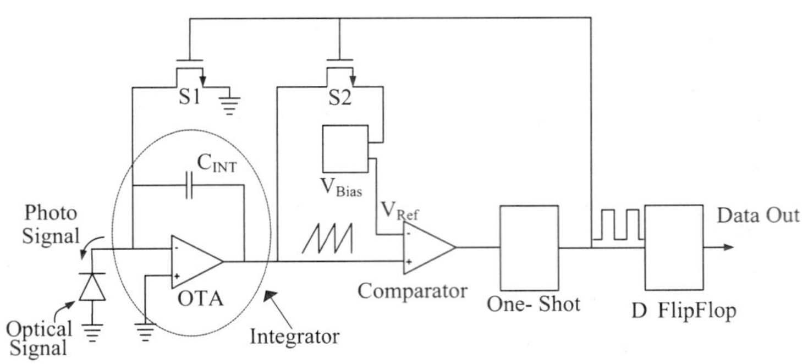

Figure 7.5 The third-generation BBIC chip. In this version, the number of MOSFET switches was reduced to two and an interior bias generation cell was added. Reproduced from Ref. [6] with permission.

voltage as its input and generates several reference voltages. The reference voltages include the comparator and the reset voltages for the OTA, which sets the initial condition on the integration capacitor before integration starts. An on-chip digital-to-analogue converter (DAC) was added for tuning out of the chip-to-chip variation. In addition, hystersis was added to the comparator to improve false triggering due to noise coupling at the input.

As with the second-generation version, having the switch in parallel with the integrating capacitor sets the OTA in a voltage follower configuration with the input at ground level, lowering the gain of the OTA prior to the integration. Similarly, the drain source voltage of S1 is set to 0 V, while the current entering the input is negligible.

7.3 Characterization of the Third-Generation BBIC

The performance of the BBIC system is highly dependent on its core building blocks. Therefore, both the photodetector and the signal-processing circuitry had to be characterized and optimized. As with the second-generation BBIC, the photodiodes are operated with a very small reverse voltage (a few millivolts), which also varies with temperature. Even though a low reverse bias increases the shot noise and the capacitance, it is still beneficial for the BBIC application as it improves the MDS value by keeping the dark currents low.

The performance of the integrator plays a crucial role in the overall system performance as it serves as the interface between the photodetector and the signal-processing circuit. The key block of the integrator is the OTA, which needs to be designed for high gain, low offset, and low flicker noise. Achieving high gain in the deep submicron process requires special circuit design techniques. The amplifier design incorporated a robust cascode scheme that was proposed by Minch [8]. The highly symmetric structure and careful layout techniques were employed to keep the offset voltage of the amplifier low. Table 7.1 summarizes the performance figures for the OTA. The layout and the photograph of the complete microluminometer chip fabricated using 0.35 μm CMOS process are shown in

Table 7.1 Key performance figures of the OTA

DC gain	92 dB
Phase margin	74°
Unity gain frequency	4 MHz
Input referred noise @ 1 KHz	110 nV/$\sqrt{\text{Hz}}$
Input referred noise @ 100 KHz	20 nV/$\sqrt{\text{Hz}}$

Reproduced from Ref. [6] with permission.

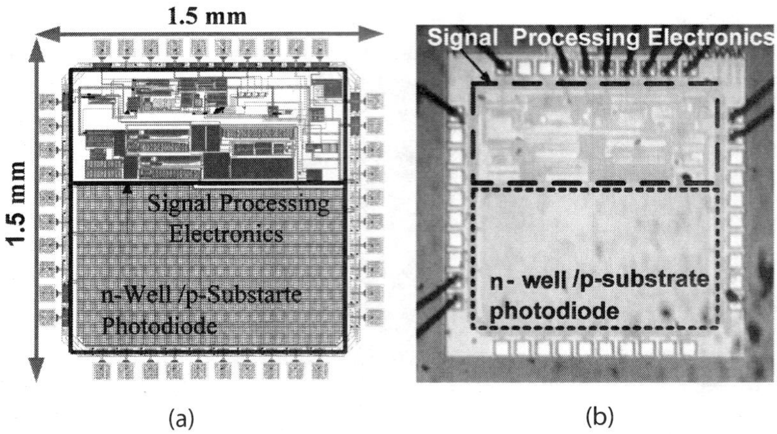

Figure 7.6 Layout and microphotograph of the microluminometer reported here. The chip measures 1500 × 1500 μm (2.25 mm^2). Reproduced from Ref. [6] with permission.

Fig. 7.6. The chip measures 1500 × 1500 μm with the photodetector occupying ~60% of the total chip area. For testing purposes the chip was mounted in a 40-pin dual in-line package (DIP).

The design enhancements made in the third generation of the BBIC system result in robustness and improved system performance when compared to the previous generations [4, 9].

7.4 Microluminometer Performance

The performance of the photodiodes in the microluminometer system is determined by the observation of the microluminometer

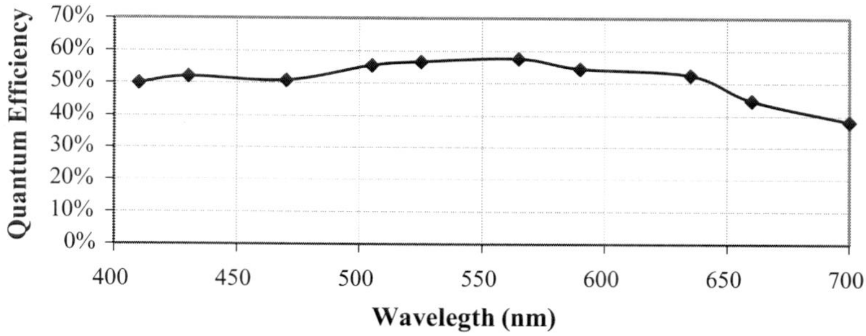

Figure 7.7 Quantum efficiency as a function of the wavelength of the incident light on photodiodes. Reproduced from Ref. [6] with permission.

output signals rather than direct probing of the photodiodes, which could produce an erroneous result due to added leakage current and parasitic capacitances. The photodiode currents were obtained by measuring the frequency of the microluminometer output pulses following Eq. 7.2. The digital logic levels and the frequency of the pulses contain the sensor data. A digital counter was employed to count the pulses for a set measurement time, which was in the range of 10–15 min. Figure 7.7 shows the measured quantum efficiency of the on-chip photodiode of the microluminometer versus wavelength of the incident light. Quantum efficiency (η) represents the effectiveness of a photodetector in converting light to a detectable current signal and is given by [4]

$$\eta = \frac{I_{\mathrm{P}}.h.\nu}{P_{\mathrm{op}}.A_{\mathrm{det}}.q}.100\% \tag{7.6}$$

where I_{p} is the photocurrent (A), h is Planck's constant (6.626×10^{-34} J-sec), υ is the optical frequency, P_{op} is the optical power (W/cm^2), A_{det} is the detector area (cm^2) and q is the charge of the electron (1.602×10^{-19} C). The quantum efficiencies of practical photodiodes are less than ideal due to the loss of excited carriers through the recombination process. The measured quantum efficiency figures of the third-generation BBIC are comparable with the previous versions developed in different process technologies [4, 9]. The bias voltage on the photodetector significantly impacts the quantum efficiency which is determined by the offset voltage of the OTA and was measured to be in the range of 8–10 mV for a V input on the positive terminal.

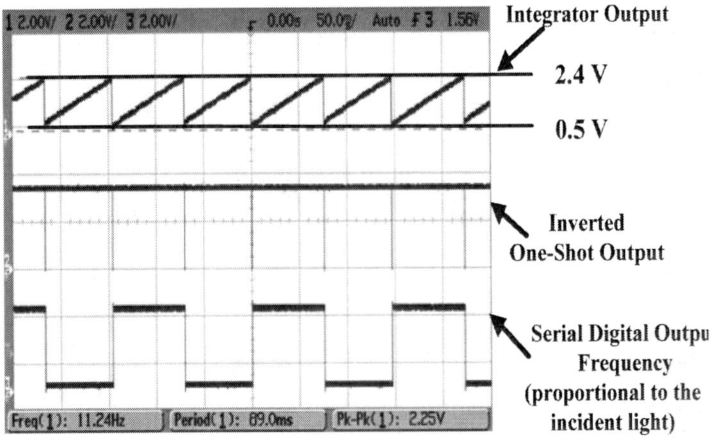

Figure 7.8 Test results of the characterization of the signal processing. Reproduced from Ref. [6] with permission.

The photodiodes were characterized using several commercial light-emitting diodes (LEDs) of various wavelengths as the light sources. The tests were performed in an integrating sphere to achieve uniformly directional light. The photodiode characteristics were measured by comparing the BBIC photocurrent to a commercial photodiode photocurrent with the same input light wavelength and intensity. At 490 nm, the quantum efficiency was observed to be about 55%.

The test results of the electrical characterization of the signal-processing component of the microluminometer with a 3.3 V power supply voltage are shown in Fig. 7.8. The combined total power consumption of the signal-processing circuitry and the photodiodes shown in Fig. 7.5 is about 3 mW. The variation in power consumption with light intensity is due to the increased frequency of operation of the one-shot and the flip-flop circuits. The frequency of each signal in Fig. 7.8 is proportional to the input light intensity component of the microluminometer using a 3.3 V power supply voltage

The leakage (dark) current was determined for several chips and found to be on the order of 40–60 fA. The variation was attributed to chip-to-chip differences introduced during the microfabrication process.

Figure 7.9 The six-chip test bed.

7.5 BBIC Instrumentation

A number of different sensor instruments have been developed using BBIC. A few of these instruments are described below.

7.5.1 *Six-Chip Test Bed*

This arrangement for chip testing contains three boards that house six microluminometers [10]. For this testing purpose, the chips are packaged in a 40-pin DIP. Each of the six microluminometers is independent and can be tested simultaneously, allowing several sets of data to be obtained under the same test conditions. When temperature, humidity, pressure, and light-level conditions are controlled and ensured to be the same, more accurate conclusions can be made about the target pollutant. This design also adds a redundancy factor to the test system, allowing for statistical analysis that can be used to locate and disregard outlying or invalid readings from the chips. The setup is shown in Fig. 7.9.

Another test was performed on the BBIC system to detect bioluminescence from bioreporters exposed to a gaseous phase

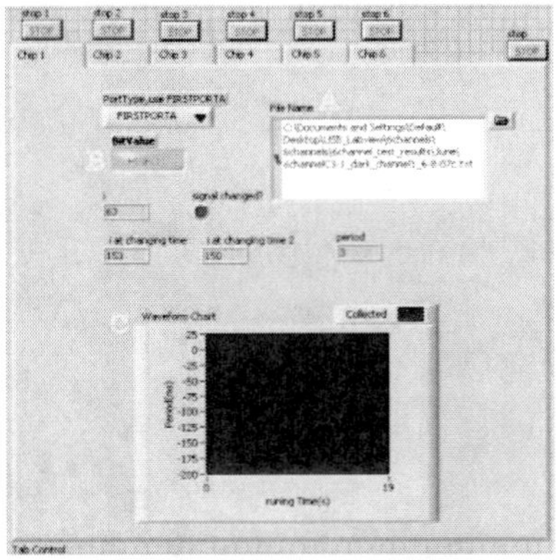

Figure 7.10 A LabVIEW .vi file. (A) The path to save data. (B) A display showing change in signal. (C) A chart to display the output of the chip.

analyte. *P. fluorescens* 5RL, a naphthalene bioreporter, was used to detect the presence of naphthalene in air. This setup uses a single USB interface (USB-1024LS) to read the data. The USB-digital input/output (I/O) module is convenient and allows testing to be performed from virtually any laptop computer without having to carry around a power supply in the field. The 24-bit digital I/O USB interface allows for high-resolution readings that can be made quickly due to the high data rate of the bus. A LabVIEW program (National Instruments, Austin, TX, USA) was used to record and plot the six distinct readings. The .vi file was simplified for six-chip testing using the tab function, as shown in Fig. 7.10.

In future BBIC systems, the USB interface will be replaced with a wireless transmitter. The transmitter will ideally send the data to a receiver that will connect with the computer in the same manner as the current USB solution. This will allow for the same software to be used with small variations made to the program. If there is a need for serial port entry for the receiver, that option will also be simple to implement into the current setup. Two possible communication standards are under consideration, Bluetooth and ZigBee technologies. Bluetooth may be ruled out due to the 10 m range restriction. ZigBee is a better choice for the intended field application of the BBIC due to the low power, low bandwidth requirements, and a range of up to 1500 m.

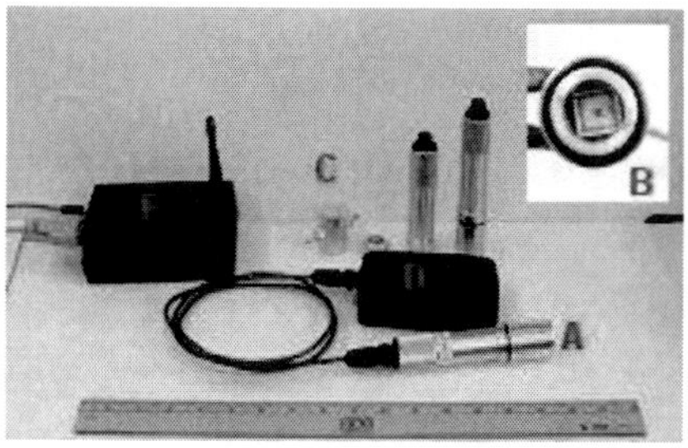

Figure 7.11 Version 1 of the BioSniffer utilized a wand-type sensor (A) with an end-on mounted microluminometer chip (B), interchangeable caps for static or flow-through monitoring (C), and a standard battery pack (D) with remote receiver (E).

7.5.2 *BBIC Wand*

The BBIC platform has been ruggedized and tailored for plug-and-play detection of desired airborne contaminants, resulting in the first-generation handheld wand-type BioSniffer sensor depicted in Fig. 7.11. This allowed for practical field testing of environmental pollutants. In future versions the entire package will be reduced to the size of the wand, with only a detachable antenna protruding from the package.

Another version of the wand has been built and tested, as shown in Fig. 7.12. The approximate dimensions of this wand are a length of 10 cm and a diameter of 3 cm. In this design the bioreporters are enclosed in a cap that is fixed on the wand tip. As the wand is moved around in air or water, the bioreporters sense the target agent and the resulting bioluminescence is detected and processed by the BBIC microluminometer. Since the wand is intended to be immersed in water during testing, the wand components and the transmitter antenna were waterproofed. This wand works essentially the same way as the six-chip arrangement. The data taken from the wand is processed using the LabVIEW program and then is analyzed. Future improvements of the wand will include wireless data transmission capability and an output LED on the exterior. Several possible LED configurations are envisioned. The simplest design will control an LED to light up when the target agent has reached a specific threshold. This will make the wand more convenient for rapid

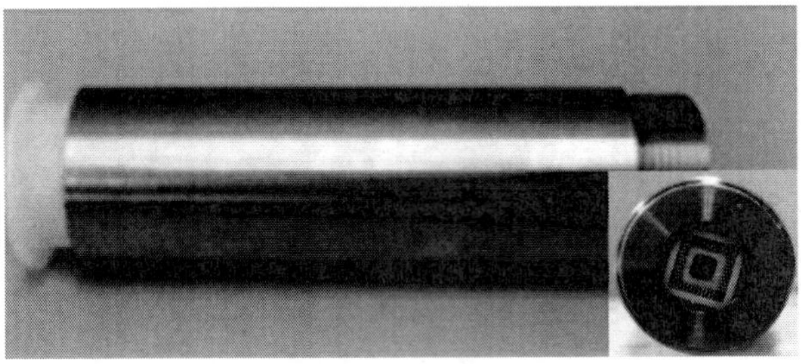

Figure 7.12 An alternate version of the BioSniffer wand. Reproduced from Ref. [10] with permission.

screening of sites. If during field testing the LED turns on then further testing will need to be performed on that site; if the LED does not turn on, then there is no need for further testing. Another option is to add a miniaturized seven-segment LED board to the wand to permit the acquisition of more useful quantitative data.

7.5.3 *BBIC CD*

A commercial handheld compact disc (CD) player was modified to include the BBIC components. With this design a rotating disk functions to distribute a single sample to multiple wells that hold different bioreporters, thus allowing for the detection of several agents simultaneously. This is a fieldable example of the six-chip test bed discussed above. Figure 7.13 shows the theoretical design

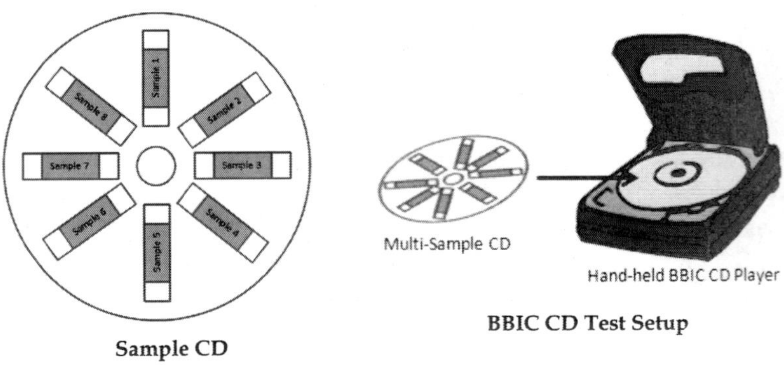

Figure 7.13 A theoretical BBIC CD.

Figure 7.14 The first BBIC CD prototype.

of the BBIC CD player and an eight-well CD sample holder. The alignment between the bioreporters and the BBIC chip is critical for a successful measurement.

A prototype handheld CD player was built containing the BBIC electronics and the controls to spin the sample disk. Figure 7.14 displays the first prototype built and tested demonstrating the BBIC CD capabilities. It was quickly discovered that meeting the light-tight constraint was not feasible using a commercial unit. However, with the capabilities of 3D printing, it would be simple to design and build a player that meets the light-tight requirement.

Another design option would be for the disk to contain only a single bioreporter in all wells to allow simultaneous testing of multiple samples. The BBIC CD provides a cheap, disposable, lightweight, and fieldable option for testing single or multiple agents in multiple samples. Early evaluation shows that additional work is required, but this design is a viable option with great potential.

(a) (b)

Figure 7.15 An air-testing chamber. (a) Side view with an air-tight lid sealed on the chamber. (b) Top view of the chamber without a lid.

7.6 BBIC Prototype Testing

The third-generation BBIC system was tested to detect bioluminescence from bioreporters exposed to gaseous- and liquid-phase analytes. For air testing of BBIC, *P. fluorescens* 5RL, a naphthalene bioreporter, was used to detect the presence of naphthalene in an air stream. *P. fluorescens* 5RL was created by inserting a transposon containing the *luxCDABE* gene cassette from *Aliivibrio fischeri* into the salicylate hydroxylase (*nahG)* gene, a component in the naphthalene metabolic pathway [11]. This fusion also allows for sensing of salicylate as it is an intermediate in the *nah* metabolic. Figure 7.15 shows the light-tight enclosure used to house the electronics and allow exposure to gases. The inlet of the enclosure was baffled to allow for indirect flow of gases, while the outlet and ports were used for electronic connections. Air was passed over naphthalene crystals, and the vapor was bubbled through water to maintain a humid environment.

P. fluorescens 5RL was grown to an OD_{546} of 0.08 ($\sim$1 $\times$ 10^8 cfu/mL) in a Luria-Bertani (LB) broth supplemented with 14 mg tetracycline/mL. A 1 mL aliquot was mixed with 1 mL liquefied LB agar, deposited on a glass microscope slide, and allowed to solidify. The slide was then placed over the BBIC chip, and uninduced background light measurements were taken for 2 h. Naphthalene-saturated air was then continuously flowed through the chamber at a rate of 200 mL/min. Figure 7.16 shows that after 40 min of

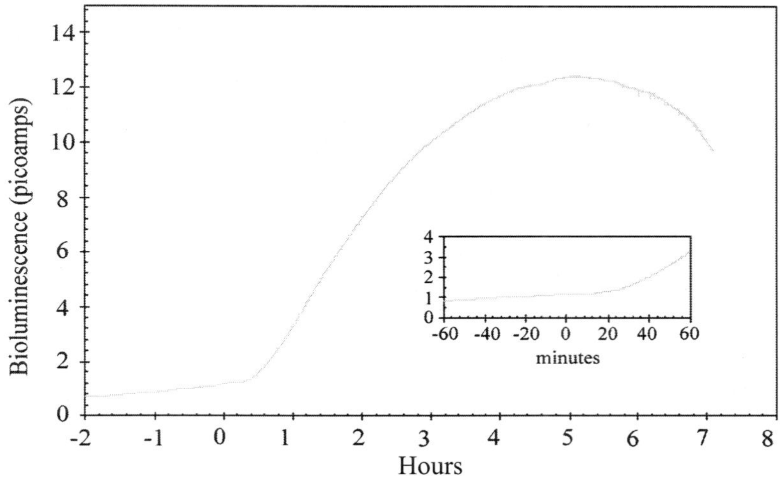

Figure 7.16 Bioluminescence of *P. fluorescens* 5RL versus time with exposure to naphthalene-saturated air in the enclosed chamber. Time points to the left of 0 correspond to data collected prior to the start of naphthalene exposure. Time 0 indicates the time at which naphthalene addition began. Reproduced from Ref. [10] with permission.

induction, a 100% increase in bioluminescence was achieved. The timescale has been normalized to display induction of the cells at time 0 h. The increase in the signal prior to induction is due to basal (noninduced) *lux* expression by the bioreporter population. After induction, the signal continues to increase for 4 to 5 h. At this point, the bioreporters become oxygen limited in the agar and reach their peak population [12]. From here the population begins to fall and bioluminescence decreases, as seen on the latter portion of the plot in Fig. 7.16.

For liquid testing of the BBIC, *P. fluorescens* 5RL was used as the bioreporter, with salicylate acting as the inducible liquid-phase analyte. Figure 7.17 shows a second light-tight enclosure designed to accommodate the microluminometer and other fixtures necessary for liquid-phase testing. The 5RL bioreporters were grown at room temperature to 1×10^8 cfu/mL in 1 L of minimal media (0.07 g KH_2PO_4, 0.17 g K_2HPO_4, 0.01 g $MgSO_4 \cdot 7H_2O$, and 0.1 g NH_4NO_3). Salicylate was then added at final concentrations ranging from

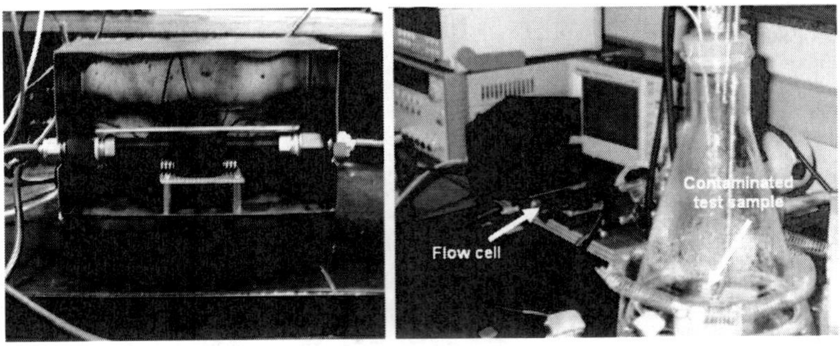

(a) (b)

Figure 7.17 Liquid testing chamber. (a) Side view of an open chamber with a flow cell covering the microluminometer. (b) The full liquid test system. Panel (b) reproduced from Ref. [6] with permission.

100 ppm to 0.05 ppm. Slow shaking of the flask (~150 rpm) created uniform bioreporter distribution throughout the solution. This solution was then circulated at a constant rate (~300 mL/min) through Viton tubing of 3 mm diameter to the clear-glass flow cell enclosure. The flow cell is situated directly over the photodiode, and the bioreporters are exposed as they flow through. At this point, salicylate was injected into the reservoir at time 0 h and the bioreporter response was recorded over a 3 h period.

Figure 7.18 shows the detection performance of the system for salicylate exposure. In this figure, the response curves for each concentration grade were normalized to the bioreporter population at the point of sample induction. The cultures began to exhibit increased bioluminescence after approximately 40 min. This was detected from samples that were drawn from the LB media and measured by spectrophotometer ($\lambda = 546$ nm). The system reached its detection limit at 0.1 ppm salicylate. The procedure used in Refs. [9, 13] was used to measure the bioluminescence.

The absolute value for the MDS with this version was calculated to be 0.3 fA, which is approximately equal to the standard deviation of the dark current. Compared to the 0.15 fA current obtained with the second version of BBIC, this value is higher due to a nonzero bias on the OTA of 10 mV.

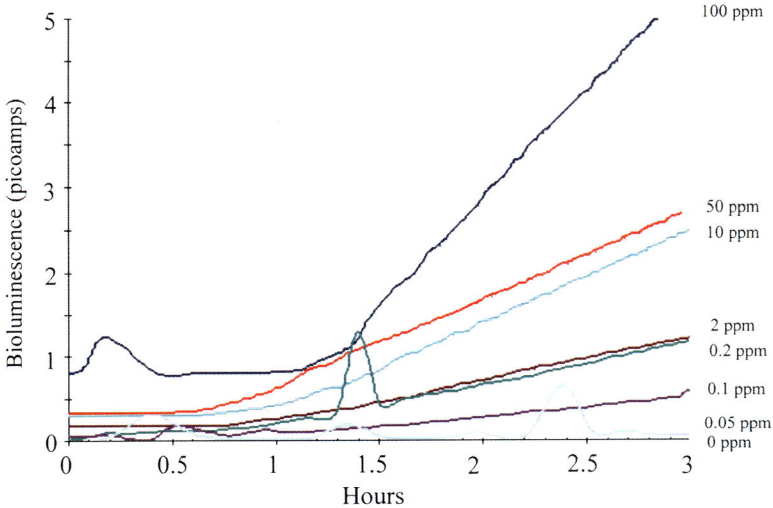

Figure 7.18 Bioluminescence of *P. fluorescens* 5RL versus time with exposure to salicylate in enclosed liquid-based test chamber. Curves display detection responses for varying concentrations of salicylate.

7.7 Conclusions

BBICs represent a biologically inspired sensing mechanism that takes advantage of self-generated *lux*-based bioluminescence for rapid, early warning monitoring of target compounds within complex environmental matrices. As the bioreporters are placed in close proximity to the microluminometer, all within a single package, these sensors thus eliminate the need for complex instrumentation to channel light from the bioreporters to the microluminometer. Since optical means are among the most flexible and prevalent mechanisms for communicating information from living cells to synthetic systems, the BBIC system reduces the cost associated with the acquisition of biologically relevant data. Because this technology can be implemented in real time within intact, living cell systems, it provides a unique and revolutionarily new perspective on bacterial, plant, and mammalian physiology and intracellular interactions. Bioluminescent assays are increasingly becoming important tools for noninvasive monitoring regimes, especially in animal model systems [3]. The monitoring of light requires less time and fewer animals than conventional methods, thus reducing the cost of obtaining biologically relevant data. With further advances in bioreporter genetics and miniaturized electronics and photonics, as well as packaging and prototyping schemes, future monitoring and detection strategies will be developed for chemical and biological agents in applications such as medical diagnostics, precision agriculture, environmental monitoring, food safety, process monitoring and control, and homeland security.

Acknowledgments

This work was supported by the National Science Foundation Division of Biological Infrastructure under award number DBI-0963854; the Division of Chemical, Bioengineering, Environmental, and Transport Systems under award number CBET-0853780; and the Army Defense University Research Instrumentation Program.

References

1. Eltzov, E., Marks, R. S. (2011). Whole-cell aquatic biosensors. *Anal. Bioanal. Chem.*, **400**(4), 895–913.
2. Su, L. A., et al. (2011). Microbial biosensors: a review. *Biosens. Bioelectron.*, **26**(5), 1788–1799.
3. Close, D. M., Ripp, S., Sayler, G. S. (2009). Reporter proteins in whole-cell optical bioreporter detection systems, biosensor integrations, and biosensing applications. *Sensors*, **9**(11), 9147–9174.
4. Bolton, E. K., et al. (2002). Integrated CMOS photodetectors and signal processing for very low-level chemical sensing with the bioluminescent bioreporter integrated circuit. *Sens. Actuators B*, **85**(1–2), 179–185.
5. Simpson, M. L., et al. (1998). Bioluminescent-bioreporter integrated circuits form novel whole-cell biosensors. *Trends Biotechnol.*, **16**(8), 332–338.
6. Islam, S. K., et al. (2007). Integrated circuit biosensors using living whole-cell bioreporters. *IEEE Trans. Circuits Syst. I*, **54**(1), 89–98.

7. Vijayaraghavan, R., et al. (2007). A bioreporter bioluminescent integrated circuit for very low-level chemical sensing in both gas and liquid environments. *Sens. Actuators B*, **123**(2), 922–928.
8. Minch, B. (2002). A low-voltage MOS cascode bias circuit for all current levels. *IEEE Int. Symp. Circuits Syst.*, **3**, 619–622.
9. Patterson, G. W. (2000). A low noise microluminometer for a bioluminescent bioreporter integrated circuit. In *Department of Electrical and Computer Engineering*, University of Tennessee, Knoxville, TN, USA.
10. Bull, N. D., et al. (2008). Genetically-engineered whole-cell bioreporters on integrated circuits for environmental monitoring. In *IEEE International Symposium on Circuits and Systems (ISCAS)*, Seattle, WA, USA.
11. King, J. M. H., et al. (1990). Rapid, sensitive bioluminescent reporter technology for naphthalene exposure and biodegradation. *Science*, **249**(4970), 778–781.
12. Nivens, D. E., et al. (2004). Bioluminescent bioreporter integrated circuits: potentially small, rugged and inexpensive whole-cell biosensors for remote environmental monitoring. *J. Appl. Microbiol.*, **96**(1), 33–46.
13. Bolton, E. K. (2001). A CMOS micorluminometer for use in a bioluminescent bioreporter integrated circuit. In *Department of Electrical and Computer Engineering*, University of Tennessee, Knoxville, TN, USA.

Chapter 8

BOD Sensor with Immobilized Luminous Cells on Chip

Toshifumi Sakaguchi[a] **and Eiichi Tamiya**[b]

[a] *Department of Environmental Sciences, Prefectural University of Hiroshima, 562, Nanatsuka, Shobara, Hiroshima 727-0023, Japan*

[b] *Division of Nanobio-Engineering, Department of Applied Physics, Graduate School of Engineering, Osaka University, 2-1, Yamadaoka, Suita, Osaka 565-0871, Japan*

sakaguchi@pu-hiroshjima.ac.jp, tamiya@ap.eng.osaka-u.ac.jp

8.1 Introduction

Rapid and onsite detection of organic pollution is an important technique for monitoring in aquatic environments and wastewater treatment. Because of its intense requirement in the bioremediation process of contaminated soil and water treatments the development of a simple and portable device for environmental assessment, which can be employed by users who do not have special skills, has became a main subject of biosensors. As the reasons are responsibility, simple operation, and cost performance, bioluminescence can be used as a reliable reporter for the assessment or monitoring of various aquatic samples containing toxicants such as pesticides,

Bioluminescent Microbial Biosensors: Design, Construction, and Implementation
Edited by Gérald Thouand and Robert S. Marks

ISBN 978-981-4613-65-1 (Hardcover), 978-981-4613-66-8 (eBook)
www.panstanford.com

polychlorinated biphenyls (PCBs), aromatic hydrocarbons, fuels, alkanes, alcohols, and heavy metals [1–16]. Nowadays, many whole-cell-based biosensors with luminescent wild-type or genetically engineered bacteria are proposed for monitoring and detection of organic pollution [17], inorganic toxicants [18], aromatic compounds [19–21], and androgenic compounds [22] in wastewater and/or environments.

Bioluminescence from a natural luminous marine bacterium, *Photobacterium phosphorous,* was also applied for the measurement of biodegradable substances [23]. Furthermore, a new rapid biochemical oxygen demand (BOD) sensing system using luminescent recombinants of *Escherichia coli* cells with *lux A–E* genes from *Vibrio fisheri* has been developed. BOD detection of multiple samples in real wastewater can be achieved by using the bioluminescence property of this recombinant *E. coli* and an imaging system containing a charge-coupled device (CCD) camera and a photomulticounter [24]. These reports show that the luminous microorganism can be used as a disposable biodevice or bioreagent in the detection of organic pollution. And they also show that a complicated cell maintenance process such as membrane immobilization of active cells is unnecessary. Additionally, recent progresses on cell-based arrays and on-chip cell assay technologies provide many possibilities of multisample, multisensing, and high-throughput analyses [25–30]. This technique is applied to the development of a mobile-type BOD measurement system for onsite detection by using a whole-cell-array-based biochip with a luminous bacterium [31, 32]. In this chapter, we describe the development of a new rapid and onsite BOD-monitoring and toxicant detection system that arrayed and immobilized luminous cells on the holes (diameter: 1000 μm) of an acrylic chip.

8.2 Isolation and Preparation of Luminous Microorganisms

Many marine organisms, fishes, and sediment samples were collected from various seashore and marine environments for the

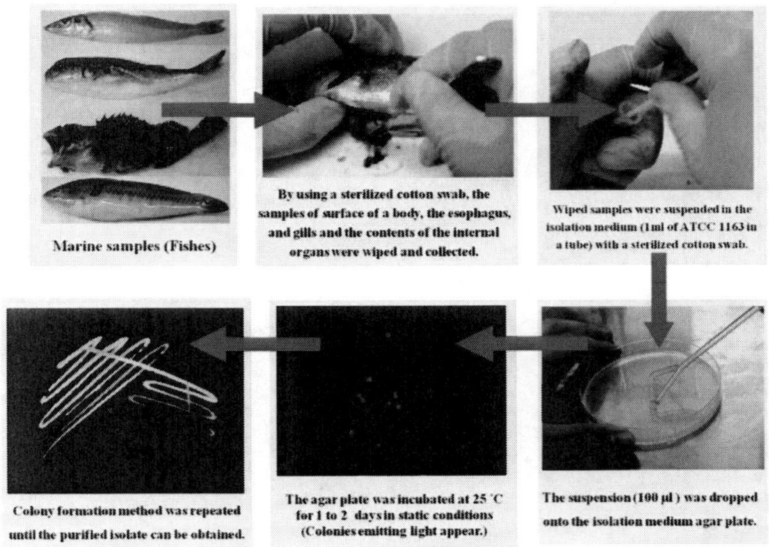

Figure 8.1 Screening procedure of the luminous bacterial isolates from marine samples (fishes).

isolation of luminescent bacteria capable of producing statically bioluminescence during the growth. Marine luminescent bacteria were grown in an ATCC 1163 medium. Many luminous isolates were obtained by the colony formation method (Fig. 8.1). After purification of luminous marine colonies, bioluminescence of the isolates was observed during the growth in the ATCC 1163 medium. Generally, bioluminescence from the culture was completely lost for the logarithmic growth phase. Several isolates capable of emitting bioluminescence after the late logarithmic growth phase were obtained. These microorganisms were isolated from marine fishes such as a rockfish, *Sebastes ventricosus* (strain MN-2); a multicolor-fin rainbowfish, *Halichoeres poecilopterus* (strain KN-5); and a Japanese jack mackerel, *Trachurus japonicas* (strain AT-1). As for these isolates, the light emission durability was higher than that of the wild type (strain IFO 13896) of *Photobacterium phosphoreum* (Fig. 8.2). They were used as a biodevice of the cell-based biochip in our developed BOD measurement system.

Figure 8.2 Culture appearance (corresponding to the late logarithmic growth phase, the OD (660 nm): approximately 0.3–0.5) and bioluminescence of the new luminous bacterial isolates, strains KN-5, MN-2, and AT-1 from marine samples (fishes). The culture was incubated in a 20 mL ATCC1163 medium (in a 50 mL grass flask) for 12 h at 25°C with shaking at 120 rpm. The initial cell concentration was prepared at 1.0×10^6 cells/mL.

8.3 Construction of an Onsite Detection System

Our newly developed onsite monitoring system features a digital camera (e.g., Fuji Film Fine Pix S602 or Canon EOS Kiss Digital N), a camera holder (chip stage), a light shade box (containing a blackout curtain or sheet), a laptop computer (containing a smart media device), and the cell-based biochip, which was immobilized cells of luminous bacteria (Fig. 8.3) [32]. The data were transferred to a (mobile) laptop computer with a smart media card device. Luminescent intensity was numerated by a black-and-white scale using Scion Image software. This system is compact and is portable in every place. Furthermore, our developed onsite detection system can measure BOD values with the free software without the purchase of a special apparatus and instruments and without requiring a high-analysis technique.

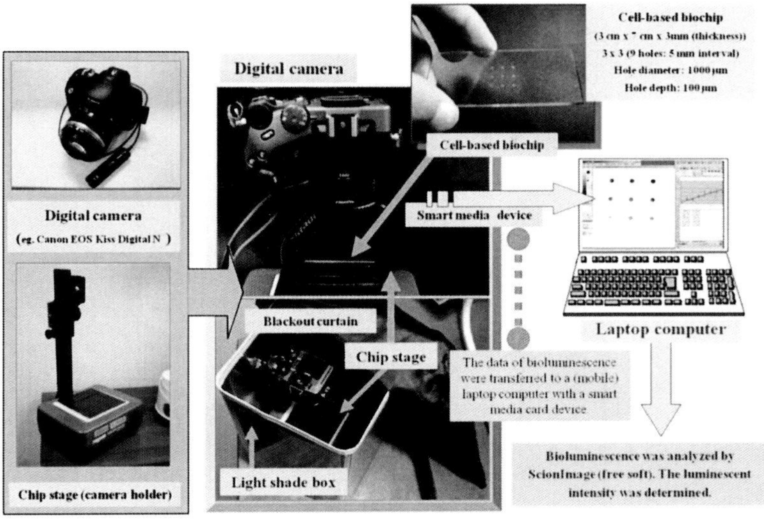

Figure 8.3 Overview and measuring procedure of an onsite BOD measurement system with a cell-based microbial biochip, a digital camera, and a laptop computer.

8.4 Biochip Preparation and Calibration

Nine holes (100 μm in depth, 1000 μm in diameter, every 5 mm) were produced on an acrylic board (3 × 7 cm) of 3 mm thickness by a micromachine processing for the manufacture of biochip. Luminescent cells (strain KN-5 grown in ATCC1163 until OD 660 nm) approximately 0.3) were immobilized with 3% (w/v) sodium alginate gel into each of the manufactured holes. For preparation of the luminous-cell-immobilized biochip with sodium alginate gel, 1% (or 0.9%) of calcium chloride solution was dropped onto the holes on the acrylic chip board (Fig. 8.4). A glucose–glutamic acid (GGA) solution (150 ppm glucose and 150 ppm glutamate) equivalent to a BOD_5 value of 220 ppm was prepared for calibration of BOD values [24]. The BOD standard solution was diluted and used for the calibration. The BOD standard solution that was treated with pure oxygen gas for 2 min was dropped onto the hole of the cell-based biochip in which were immobilized luminous cells of marine bacterial isolates with 3% (w/v) sodium alginate

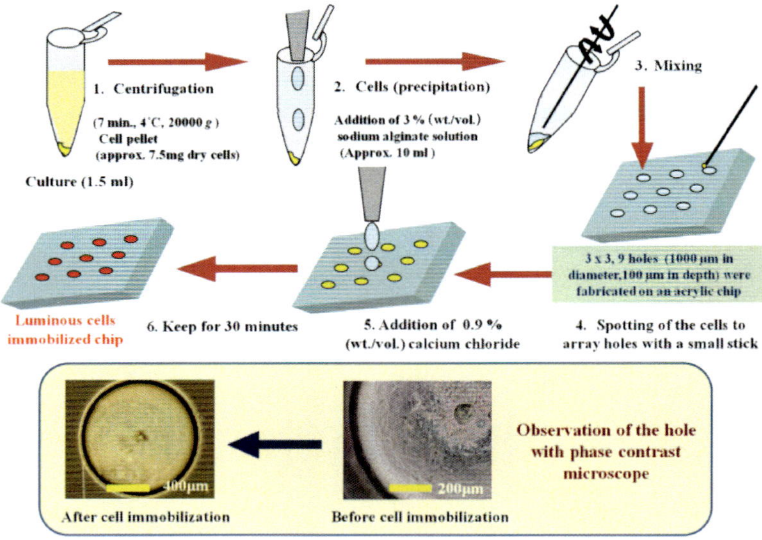

Figure 8.4 Procedure of luminous cell-based biochip preparation and observation of fabricated hole on the biochip with the phase contrast microscope.

gel. After incubation at room temperature (about 25°C) for 20 min bioluminescence from the holes on the biochip was observed and the photointensity was estimated by gray-scale images with Scion Image software. The calibration curve for BOD determination was evaluated by comparison with that of the chemi-imager system (a chemiluminescence detector (Aisin LumiVision Pro HS II).

Cells (strain KN5: A and B and *Photobacterium phosphoreum* strain IFO 13896: C) were incubated in the same conditions described in Fig. 8.2. Grown cells were immobilized by the same manner shown in Fig. 8.4.

An almost similar luminescent image was observed with our onsite system compared to the case of a chemi-imager (Fig. 8.5). A clear image of bioluminescence from holes on the biochip was obtained by using our developed system in the almost same revel of the chemi-imager system. Furthermore, the chip images demonstrated that this system allowed us to visualize simultaneously different standard solutions. BOD values of the multiple standard samples could be determined by using this chip system in a single, small

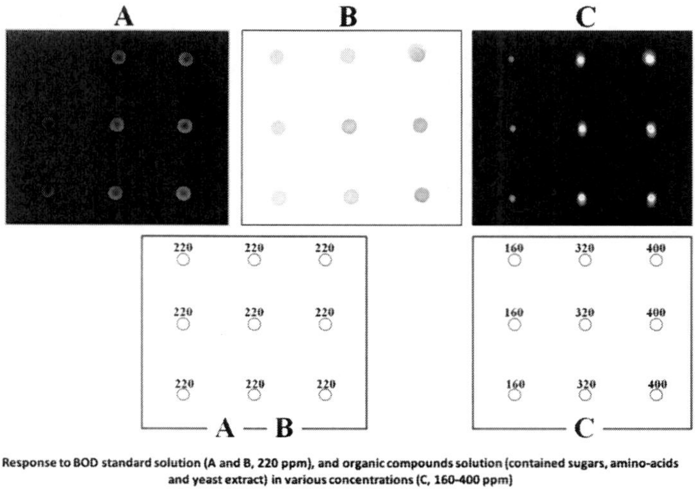

Figure 8.5 Comparison of bioluminescent image among our developed onsite BOD measurement system (the direct image (A) and the gray scale image (B)) and the chemi-imager system (Aisin, LumiVision Pro HS II) (C).

acrylic plate. Rapid determination of the calibration curve has been achieved by using this chip system. A similar level of calibration curve and equation with the case of a chemi-imager system could be plotted and calculated (Fig. 8.6). On the basis of a linear relationship between photointensity (bioluminescence) and BOD concentration below 16 ppm, a linear calibration curve and equation (our system and the chemi-imager system, respectively) were estimated and approximated from the calibration curve for determination of BOD values in actual samples (Fig. 8.6).

8.5 Practical Application and Toxicant Detection

Measurement of organic pollution due to biodegradable substances in actual samples from the wastewater treatment plant in a confectionery factory was performed by using our chip system. The images and obtained values demonstrated that this system also allowed us to visualize simultaneously the change of organic pollution in each treatment step (Table 8.1). Moreover, BOD values

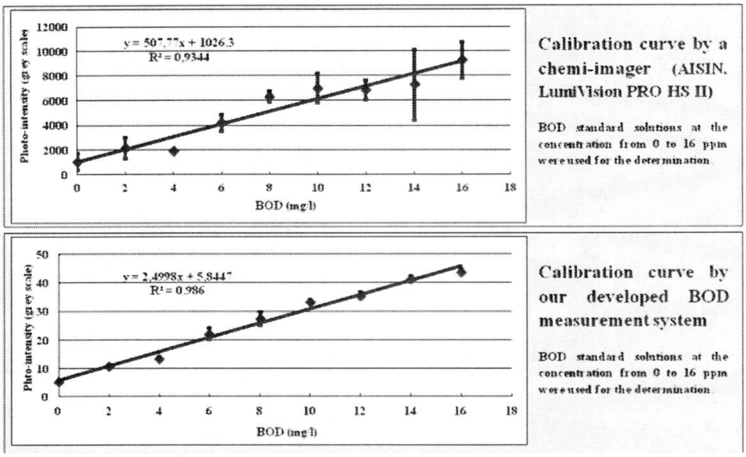

Figure 8.6 Calibration curves by the chemi-imager system (a chemiluminescence detector, Aisin, LumiVision Pro II) and our developed onsite BOD measurement system.

Table 8.1 Comparison of BOD values that were determined by using the onsite measurement system with those measured by the 5-day method for actual samples from the wastewater treatment plant in a confectionery company

Sampling place	Onsite measurement system (ppm)	BOD_5 (JIS method) (ppm)
River (Fukuokapref.)	4.85 ±0.80	5.25
Treated water (water treatment plant)	5.94 ± 0.91	6
Consolidated tank water (water treatment plant)	298.76 ± 11.29	315
Oil trap tank water (water treatment plant)	503.46 ±34.42	491
Aeration tank water (water treatment plant)	294.51 ±13.78	271

of the multiple actual samples could be determined by using this chip system in a single, small acrylic plate. A clear correlation could be observed between the two values (our system and BOD_5). The difference in BOD values was within 30%. The values were almost corresponding. These results demonstrated that almost the same performance as the conventional method (BOD_5) was admitted for the determination of BOD values in actual samples. Moreover, our system could be applied to detect cationic and anionic toxicants on

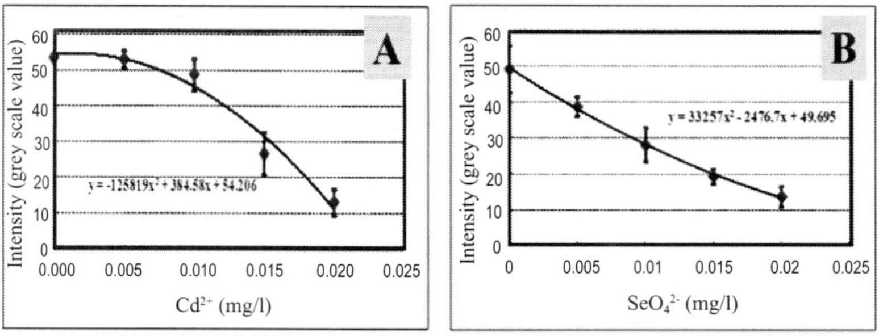

Figure 8.7 Toxic cation (A) cadmium chloride and anion (B) sodium selenate detection due to the inhibition of bioluminescence. Toxicity was detected and estimated by the decrease of bioluminescence when a sample toxic ion solution was added to a 16 ppm BOD standard solution. The photointensity was compared with that of the original 16 ppm BOD standard solution.

the basis of the inhibition of bioluminescence (Fig. 8.7). Toxic heavy metal cations and harmful anions such as silver, nickel, manganese, zinc, cobalt, copper, cadmium, selenate, selenite, tellurate, tellurite, arsenate, arsenite, dichromate, and borate could be detected by the inhibition of bioluminescence when these ions existed in the sample solution as sole ion species. In the case of cadmium ions or selenate ions, the bioluminescence was drastically decreased to approximately 30% at a concentration of 0.02 mg/L (Fig. 8.7). This result showed that highly sensitive toxic ion detection was achieved by using our developed system.

8.6 Conclusions

In this chapter, we have introduced an imaging system containing a commercial-based digital (CCD) camera, a mobile PC, and luminous photobacterial cells immobilized on an acrylic chip for BOD measurement. Results showed that it is possible to determine the degree of pollution due to the presence of biodegradable organic substances in multiple samples by using our luminous cell-based biochip system within 20 min using one drop of sample. This system had an almost same performance as the chemi-imager system for the detection of BOD in environmental samples. Moreover, our system does not need high-cost experimental instruments and apparatus. It is portable and can be easily used on pollution sites and wastewater treatment sites. BOD values lower than 16 ppm can be directly detected without dilution of the sample onsite. BOD values more than 16 ppm also can be rapidly measured by diluting the sample onsite. Furthermore, our system could be used to estimate toxicity of heavy metal cations (e.g., cadmium, copper, and manganese) and harmful anions (e.g., selenate and arsenate) by the inhibition of bioluminescence. These results suggest that the onsite and high-throughput determination of BOD and the detection of toxicant can be achieved. In future, our biochip system that used bioluminescence will be also a powerful technique to detect and monitor specific compounds in onsite wastewater by using a gene promoter assay with recombinant cells that contain luminescence genes from luminous organisms.

Acknowledgments

This work was partly suported by Grant-in-Aid on research theme search for industrial promotion from Iizuka City in Fukuoka, Japan (2001). We also thank the Micro-Space Chemistry Laboratory in AIST Kyushu for technical support and advice on microfabrication. TS also deeply acknowledges graduates and students of Sakaguchi Laboratory who participated in and contributed to this study, Mr. Hiroaki Mizoguchi, Mr. Yasunori Morioka, Mr. Masahiro Yamasaki, Mr. Junpei Iwanaga, and Mr. Kazuhiko Beppu. We gratefully thank Grant-in-Aid for Scientific Research (C) Japan (subject number 19510011) for financial support.

References

1. Van Dyk, T. K., Smulski, D. R., Reed, T. R., Belkin, S., Vollmer, A. C., LaRossa, R. A. (1995). *Appl. Environ. Microbiol.*, **61**, 4124–4127.
2. Weitz, H. J., Campbell, C. C., Killham, K. (2002). *Environ. Microbiol.*, **4**, 422–429.

3. Ulitzur, S., Lahav, T., Ulitzur, N. (2002). *Environ. Toxicol.*, **17**, 291–296.
4. Backhaus, T., Grimme, L. H. (1999). *Chemosphere*, **38**, 3291–3301.
5. Ruiz, M. J., Lopez-Jaramillo, L., Redondo, M. J., Font, G. (1997). *Bull. Environ. Contam. Toxicol.*, **59**, 619–625.
6. King, J. M., Digrazia, P. M., Applegate, B., Burlage, R., Sanseverino, J., Dunbar, P., Larimer, F., Sayler, G. S. (1990). *Science*, **249**, 778–781.
7. Add-El-Haleem, D., Ripp, S., Scott, C., Sayler, G. S. (2002). *J. Ind. Microbiol. Biotechnol.*, **29**, 233–237.
8. Stocker, J., Balluch, D., Gsell, M., Harms, H., Feliciano, J., Daunert, S., Malik, K. A., Van de Meer, J. R. (2003). *Environ. Sci. Technol.*, **37**, 4730–4750.
9. Choi, S. H., Gu, M. B. (2002). *Biosens. Bioelectron.*, **17**, 433–440.
10. Cho, J.-C., Park, K.-J., Ihn, H.-S., Park, J.-E., Kim, S.-Y., Kang, I., Lee, K.-H., Jahng, D., Lee, D.-H., Kim, S.-J. (2004). *Biosens. Bioelectron.*, **20**, 338–344.
11. Ren, S., Frymier, P. D. (2003). *Water Environ. Res.*, **75**, 21–29.
12. Tauriainen, S., Krap, M., Chang, W., Virta, M. (1998). *Biosens. Bioelectron.*, **13**, 931–938.
13. Minak-Bernero, V., Bare, R. E., Haith, C. E., Grossman, M. J. (2004). *Biotechnol. Bioeng.*, **87**, 170–177.
14. Kobatake, E., Niimi, T., Haruyama, T., Ikariyama, Y., Aizawa, M. (1995). *Biosens. Bioelectron.*, **10**, 601–605.
15. Eltzov, E., Marks, R. S. (2011). *Anal. Bioanal. Chem.*, **400**, 895–913.
16. Durand, M. J., Thouand, G., Dancheva-Ivanova, T., Vachon, P., DuBow, M. (2003). *Chemosphere*, **52**, 103–111.
17. Olaniran, A. O., Motebejane, R. M., Pillay, B. (2008). *J. Environ. Monit.*, **10**, 889–893.
18. Olaniran, A. O., Hiralal, L., Pillay, B. (2011). *J. Environ. Monit.*, **13**, 2914–2920.
19. Behzadian, F., Barjeste, H., Hosseinkhani, S., Zarei, A. R. (2011). *Curr. Microbiol.*, **62**, 690–696.
20. Kim, M. N., Park. H. H., Lim, W. K., Shin, H. J. (2005). *J. Microbiol. Methods*, **60**, 235–245.
21. de Las Heras, A., de Lorenzo, V. (2012). *Methods Mol. Biol.* **834**, 261–281.
22. Roda, A., Cevenini, L., Michelini, E., Branchini, B. R. (2011). *Biosens. Bioelectron.*, **26**, 3647–3653.
23. Hyun, C.-K., Tamiya, E., Takeuchi, T., Karube, I., Inoue, N. (1993). *Biotechnol. Bioeng.*, **41**, 1107–1111.
24. Sakaguchi, T., Kitagawa, K., Ando, T., Murakami, Y., Morita, Y., Yamamura, A., Yokoyama, K., Tamiya, E. (2003). *Biosens. Bioelectron.*, **19**, 115–121.

25. Matsui, M., Kaya, T., Nagamine, K., Yasukawa, T., Shiku, H., Matsue, T. (2006). *Biosens. Bioelectron.*, **21**, 1202–1209.
26. Nagamine, K., Matsui, N., Kaya, T., Yasukawa, T., Shiku, H., Nakayama, T., Nishino, T., Matsue, T. (2005). *Biosens. Bioelectron.*, **21**, 145–151.
27. Mishra, N. N., Retterer, S., Zieziulewicz, T. J., Isaacson, M., Szarowski, D., Mousseau, D. E., Lawrence, D. A., Turner, J. N. (2005). *Biosens. Bioelectron.*, **21**, 696–704.
28. Lee, J. H., Mitchell, R. J., Kim, B. C., Cullen, D. C., Gu, M. B. (2005). *Biosens. Bioelectron.*, **21**, 500–507.
29. Tsai, H.-C., Doong, R.-A. (2005). *Biosens. Bioelectron.*, **20**, 1796–1804.
30. Fesenko, D. O., Nasedkina, T. V., Prokopenko, D. V., Mirzabekov, A. (2005). *Biosens. Bioelectron.*, **20**, 1860–1865.
31. Sakaguchi, T., Morioka, Y., Tamiya, E. (2004). Bioluminescence and chemiluminescence, progress and perspectives. In *Proceedings of the 13th International Symposium*, Tsuji, A., Matsumoto, M., Maeda, M., Kricka, L. J., Stanley, P. E. (eds.), World Scientific, pp. 437–440.
32. Sakaguchi, T., Morioka Y., Yamasaki M., Iwanaga, J., Kazuhiko Beppu, K., Maeda, H., Morita, Y., Tamiya E. (2007). *Biosens. Bioelectron.*, **22**, 1345–1350.

Chapter 9

Bioluminescent Whole-Cell Sensors: Integration of Bacterial Reporter Cells onto Hardware Platforms

Sahar Melamed and Shimshon Belkin

Department of Plant and Environmental Sciences,
Alexander Silberman Institute of Life Sciences,
Hebrew University of Jerusalem, Jerusalem 91904, Israel
shimshon.belkin@mail.huji.ac.il

9.1 Introduction

Genetically engineered reporter bacteria have been developed over the past 30 years for the detection of diverse compounds in water, air soil and other matrices [1]. Reporter bacteria are molecularly modified to generate a dose-dependent quantifiable signal, emitted by a reporter protein, when exposed to predetermined biologically active stimuli (a chemical, a class of chemicals, or a stress factor). Such sensor bacteria hold great promise for future monitoring purposes due to their simplicity, low cost and facilitated genetic tailoring to specific applications. The system is usually engineered by fusing a sensor element, for example a gene promoter sequence

Bioluminescent Microbial Biosensors: Design, Construction, and Implementation
Edited by Gérald Thouand and Robert S. Marks

ISBN 978-981-4613-65-1 (Hardcover), 978-981-4613-66-8 (eBook)
www.panstanford.com

anticipated to be induced by the target stimulus, to a reporter gene(s) [2–4]. One of the most commonly used reporter systems is the bacterial bioluminescence genes *luxCDABE* from *Aliivibrio fischeri, Photorhabdus luminescens* or other luminescent bacteria [5, 6]. The use of the *luxCDABE* genes is advantageous in that it alleviates the need for external substrate addition.

Recent convergence of the biological, chemical, and engineering sciences has opened up new avenues of research on integrating such reporter cells with inanimate platforms. A variety of such devices has been described in the course of the last decade, some of which have been covered in recent papers and reviews [7–13]. In the present chapter we will focus on recent progress in the incorporation of bioluminescent bacterial reporters into biosensor devices. We will review the main advances in deposition and immobilization techniques and discuss the most promising means to preserve the viability and functionality of the incorporated cells.

9.2 Live Cell Patterning and Deposition Techniques

9.2.1 *Inanimate Surfaces for Live Cell Deposition*

The nature of the substratum used for cell deposition or patterning is crucial, as it may affect not only the initial integration of the live cells but also their subsequent viability and activity. The three most commonly used biochip materials are silicon, glass, and diverse soft polymers.

Silicon is the most common material in microfabrication. In general, it is characterized by chemical homogeneity and stability, manageable surface properties (such as polarity and wettability), excellent electrical properties, amenability to chemical modification with diverse functionalities, and zero autofluorescence [14, 15]. Moreover, refined and precise micromechanical structures can be designed on the basis of silicon's exceptional mechanical properties [16]. However, strong adhesion forces between the silicon surface and the biological matter may affect biosensor functionally and therefore should be taken into consideration. For example, oxidized silicon surfaces are hydrophilic, and when exposed to air they can absorb several layers of water, favoring the formation of meniscus and capillary action. A main drawback in the use of silicon for bioluminescence assays is its opacity.

Glass, also a popular biochip material, is often selected because of its total transparency and established surface chemistry, allowing the integration of biological material [17, 18]. Fused silica, the purest form of silicon dioxide (SiO_2), has low autofluorescence, has high resistance to solvents and nonfluorinated acids, can withstand high temperatures, and is amenable to shaping. Borosilicate, composed of silica and boron oxide, is less preferable in view of its higher autofluorescence.

A third option for biochip surfaces is polymeric materials, the use of which is increasing rapidly [19–23]. Their main advantages lie in their high flexibility, easy formability, and low cost, the last being an essential parameter for mass production. Yet, they have some key limitations such as low stability, autofluorescence of some polymers, poor temperature resistance, and occasionally unfavorable chemical characteristics. For instance, the hydrophobicity of one of the main polymers used, poly(dimethyl siloxane) (PDMS), poses a problem when using aqueous solutions and promotes undesirable absorption of organic solvents and hydrophobic compounds. Many of these issues may be overcome by chemical modifications of the polymers' surfaces, rendering them more compatible as biochip materials [24–26].

9.2.2 *Soft Lithography*

Cell deposition or patterning on the aforementioned surfaces can be carried out in various ways. One approach for cell patterning is to deposit the cells onto a prepatterned surface, providing a "map" in which surfaces designated for cell integration are defined in advance. A well-developed family of techniques for surface patterning is soft lithography [27–30], originally developed as an alternative for photolithography, the main workhorse of microfabrication [31]. The main concept of soft lithography is the use of an elastomeric stamp (usually made of PDMS) with relief patterns on its surface for printing, molding and embossing. These patterning techniques can generate structures from the nanoscale to the microscale, on flexible and soft substrates, either planar or curved. Soft lithography is low cost, eliminates the need for clean rooms (except for elastomeric stamp fabrication), and is friendlier for biologists since the strong photolithography solvents are avoided. The main techniques in this family are microcontact printing (μ CP) [32] replica molding (REM) [33], microtransfer molding (μ TM) [34], micromolding in capillaries (MIMIC) [35], solvent-assisted micromolding (SAMIM) [36], phase-shifting edge lithography [37], nanotransfer printing [38, 39], decal transfer lithography [40], and nanoskiving [41, 42]

9.2.3 *Cell Printing*

A different approach for cell patterning is to generate the pattern by the cell deposition technique; such an approach also allows the deposition of different reporter strains in different locations on the same platform. A recent technique demonstrated for patterning live cells is cell printing by inkjets or related noncontact technologies of high resolution [43–46]. A simple piezoelectric inkjet printer was adapted for patterning bacteria onto substrates such as a glass slide, agar plate, or nitrocellulose membrane. Arrays composed of two *Escherichia coli* strains were formed and 98.5% printed bacteria viability was determined (Fig. 9.1a,b) [47]. Melamed et al. [48] used a modified noncontact robotic printer (sciFlexarrayer S3, Scienion AG, Germany), originally designed for arraying biomolecules, to generate printed arrays of fluorescent and bioluminescent reporter cells; the printer has the ability to print ultralow liquid volumes starting from 0.1 nL per drop with a printing resolution of 1 μm. Samples (0.5 to 2.5 nL, spot size 50 to 200 μm) of bioluminescent and fluorescent bacterial reporter strains were deposited on either bare or 3-aminopropyl-triethoxysilane (APTES)-coated glass slides (Fig. 9.1c). The deposited cells retained their viability and activity for at least 2 months at 4°C.

Two of the main advantages of noncontact printing are superior spot morphologies and a reduced probability of contamination, as the substrate and the dispensing capillary are separated. The method is limited, however, by the viscosity of the printed solution

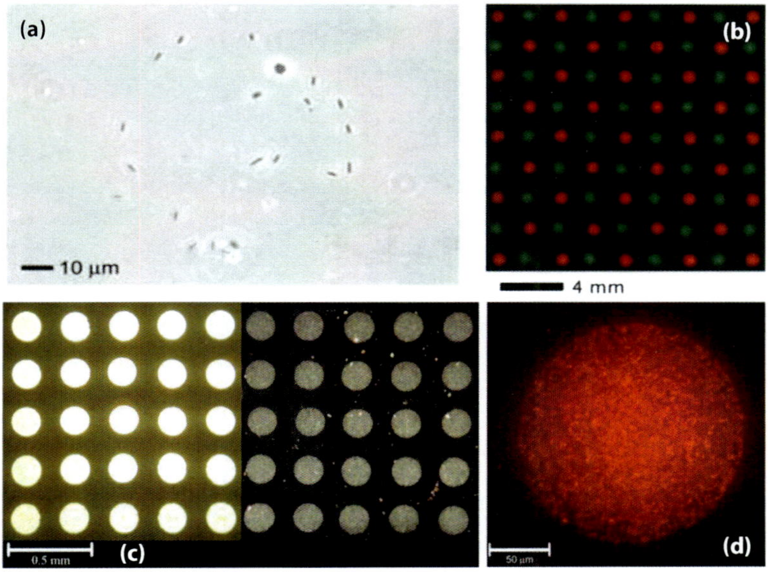

Figure 9.1 (a) One drop of inkjet deposited *E. coli* immediately after printing and (b) colony array grown after printing two types of fluorescently labeled cells (10 h at 30°C following printing); CFP-labeled cells in (pseudocolor) red and Venus-labeled cells in (pseudocolor) green [47]. Panels a and b are reprinted from Ref. [47] with permission from *PLOS ONE*. (c) *E. coli* reporter strain SM111 (*recA::DsRed2*, 1 nL) printed on APTES-coated glass immediately after printing (5 left columns) and after 24 h incubation at 4°C, following removal of nonadhered cells (5 right columns and (d)) [48]. Panels c and d are reproduced from Ref. [48] with permission from the Royal Society of Chemistry.

which may hinder deposition by a nozzled instrument, thus making it less favored for printing polymer suspensions such as alginate. This was at least partially overcome [48] by equipping a printer with a special holder that allowed raising the temperature of the printed samples up to 50°C to reduce sample viscosity. In this case the capillaries were also coated with a hydrophobic polymer in order to address another concern associated with cell printing: physical forces, such as shear rate, which may exert mechanical stress and affect cell viability and activity [49, 50].

9.3 Cell Immobilization

Integrating live cells onto a biosensor surface may be achieved either by direct attachment of naked cells or by embedding them in a suitable polymer. Successful direct attachment should be based on the compatibility of the animate and inanimate surfaces, which may be improved by either chemical modification of the inanimate substrate and/or biological manipulation of the external properties of the cells [48, 51–55]. Polymer-based immobilization offers additional advantages including high short-term viability [56–58]. Technological aspects of these approaches as employed for the immobilization of bioluminescent reporters are discussed below.

9.3.1 *Naked Attachment: Animate and Inanimate Surface Modifications*

A variety of techniques may be used to adjust the chip's surface properties for bacterial attachment. As bacterial cells are usually negatively charged [59], the use of positively charged surfaces to enhance adhesion is often promoted. For example, Choi et al. immobilized bacteria in specific patterns using positively charged self-assembled polyelectrolyte (PEL) multilayers separated by the repulsive poly(ethylene glycol)-poly(D,L-lactide) (PEG-PLA) diblock copolymer The modified surface was formed using PDMS molds with micro or nanostructures, which were placed on PEL-coated glass followed by the addition of PEG-PLA for filling the void spaces [60]. In another study, glass slides were coated with the positively charged APTES in order to promote the adhesion of the negatively charged bacterial fluorescent or bioluminescent reporters [48]. For the preparation of APTES-coated glass slides, the slides were first cleaned with acid piranha and then incubated in 1% APTES in dry toluene. Bacterial drops, in the range of 0.5 nL to 500 nL, were printed onto the glass slides using a modified noncontact robotic printer and the printed bacteria were kept at 4°C. Adhesion was analyzed using epifluorescence microscopy (filter sets 13 and 20, Axivert 135TV, Zeiss, Germany) following the removal of the nonadherent cells (Fig. 9.1c,d). When immobilizing nanoliter-scale drops of bacteria, it should be taken into consideration that the cells are dehydrated almost immediately after deposition. This can be lethal if preventive measures are not undertaken, such as the use of osmoprotectants, both extracellular and intracellular. Inside the cell, these compounds equalize the osmotic pressure and stabilize cell components, while extracellularly they shield the bacterium from the harmful effects of dehydration [61–63].

For estimating the viability and activity of the printed reporter bacteria, bioluminescent bacteria were printed onto a standard 96-well microtiter plate or onto a 96-well microtiter plate, the bottom of which was replaced by APTES-coated glass. In the latter case, the bottom of the glass slide was sealed with a white tape to make the surface more opaque (Fig. 9.2a).

In the example presented in Fig. 9.2b, the viability of constitutively bioluminescent (*CP25::luxCDABE*) reporter cells' 500 nL drops, 24 h after deposition, is compared, without treatment (M63), with intracellular ectoine or with a combination of intracellular ectoine with one of three extracellular osmoprotectants.

Using dip-pen nanolithography, Nyamjav et al. [55] likewise used electrostatic interactions between the surface, poly-L-lysine (PLL), and bacteria to generate a patterned selective surface for the adhesion of bacterial cells. In another example, *Staphylococcus aureus* cells were immobilized onto polished silicon wafers modified with thin layers of alkylsilanes exhibiting different functional groups. Oxidized surfaces and surfaces terminated either with *n*-octyltriethoxysilane or (heptadecafluoro-1,1,2,2-tetrahydrodecyl) trichlorosilane supported the immobilization of the cells, while surfaces terminated with *N*-(triethoxysilylpropyl)-*O*-poly(ethylene glycol)urethane were inhibitive [54]. Higher immobilization specificity than that displayed by electrostatic attraction may be achieved by the use of antibodies. For instance, *Salmonella enterica* cells were immobilized onto silicon and gold by using the interaction between overexpressed CFA/I fimbriae and the corresponding antibody (Fig. 9.3a,b). First, the surface was coated with PEG to prevent nonspecific binding of antibodies, and then patterns of the bare surface were generated in the PEG layer, allowing covalent bonding of the antibodies [64].

As a complementary approach for cell immobilization, modification of the bacterial cell can be performed. Bacterial adhesion

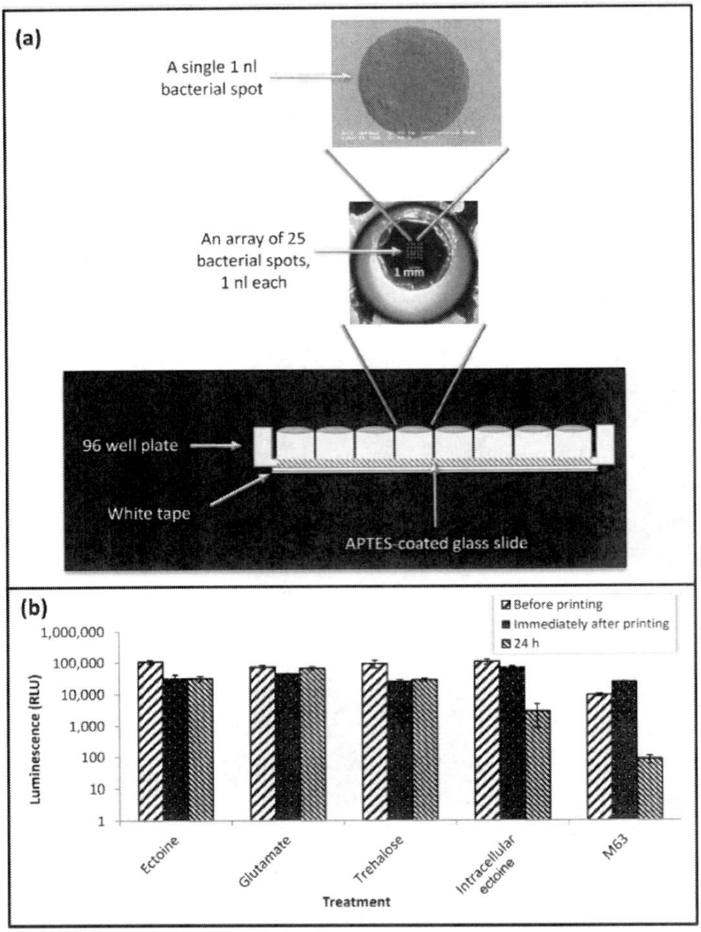

Figure 9.2 (a) Schematic drawing of a 96-well plate, the bottom of which was replaced by APTES-coated glass (bottom). Array of 1 nL reporter bacteria drops in one well (middle) and one bacterial spot (top) is shown (high-resolution SEM). (b) Residual activity of a constitutively bioluminescent (*CP25::luxCDABE*) *E. coli* reporter strain SM117 following desiccation and rehydration. Spots (500 nL) were printed either with no addition (M63), pregrowth under osmotic stress (NaCl 0.5 M) and ectoine (Intracellular ectoine) or with pregrowth under osmotic stress (NaCl 0.5 M) and ectoine with different osmoprotectants added to the printing medium. Bioluminescence is displayed before printing, immediately after printing, and after incubation for 24 h at 4°C. Panels a and b are modified from Ref. [48] with permission from the Royal Society of Chemistry.

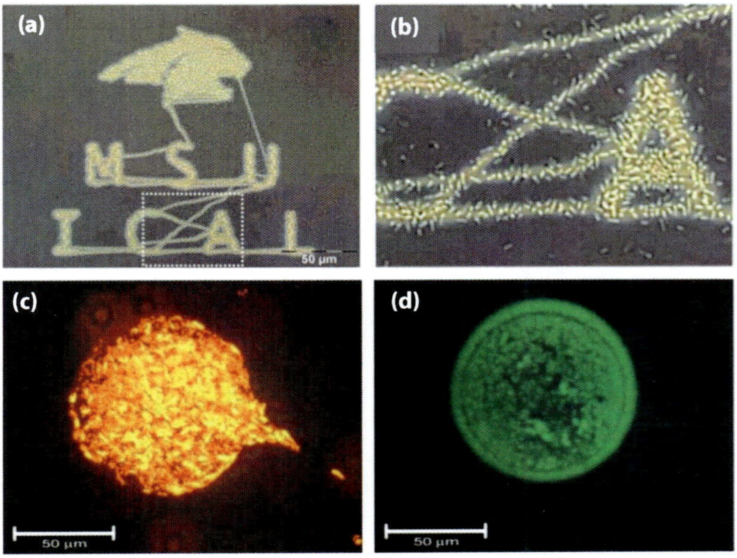

Figure 9.3 (a) Bobcat mascot and MSU-ICAL (Montana State University–Imaging and Chemical Analysis Laboratory) logo of live *S. enterica* cells immobilized on silicon etched by a focused ion beam; (b) enlargement of the area within the white-dotted box in (a), demonstrating the cellular resolution of cell patterning [64]. Panels a and b reprinted with permission from Ref. [64]. Copyright (2008) American Chemical Society. (c) Overproducing curli fluorescent (*recA::DsRed2) E. coli* cells adhered to PLL spot on PEG-coated glass following a 24-hour incubation in minimal medium; (d) the same field as in (c), showing the PLL spot, which was prelabeled with fluorescein isothiocyanate (Melamed et al., unpublished).

may be stabilized by natural outer-cell surface structures, the manipulation of which can be used to modify adhesion intensity. Pili, one of the main surface structures, are filamentous appendages that extend from both gram-negative and gram-positive bacterial cells, mediating adhesion to surfaces and to host cells, as well as biofilm formation [65–68]. For example, mutations in FimH, one of the subunits of type 1 pili in *E. coli* [69], can enhance adhesion to surfaces [52]. Melamed et al. [48] used an *E. coli* strain that overproduces curli, an extracellular surface structure that forms an adhesive extracellular matrix, to promote the adhesion to glass coated with PLL (Fig. 9.3c,d) (Melamed et al., unpublished) or APTES

[48]. It is also possible to use more than one factor to enhance bacterial adhesion, as was done by exploiting the synergistic effect between curli and cellulose [70]. Additional information on different techniques for manipulating cell–surface interactions is provided in a recent review by Colpo et al. [71].

9.3.2 *Immobilization by Deposition of Encapsulated Cells*

A different approach for cell immobilization is the encapsulation of the cells inside hydrogels, flexible hydrophilic polymers with high water content.

Entrapping bacteria in a hydrogel is simple and rapid, but care should be taken to avoid deleterious effects on cell viability and activity, caused either by the polymer components or by the immobilization protocol. In the last decade a variety of hydrogels such as polyacrylamide, sodium alginate, sol-gels, agarose, PEG-based hydrogels and agar were used to encapsulate bacteria [56, 58, 72–75]. Detailed reviews on the advantages and disadvantages of different hydrogels for cell encapsulation can be found elsewhere [13, 76]. In this section we will provide a few up-to-date examples of bioluminescent strain encapsulation. Elad et al. [58] constructed a flow-through biosensor for online continuous water toxicity monitoring. The biosensor consisted of four disposable PDMS biochips harboring agar-immobilized bioluminescent reporter bacteria, forming an array of 12 bacterial spots in each biochip. The use of agar for cell encapsulation exhibited advantages such as biocompatibility, high stability, low cost, and a simple preparation protocol. The responses of the reporters to drinking water flowing above them were probed by single-photon avalanche diode detectors (Fig. 9.4a). Using a panel of four reporter strains, the successful concomitant detection of three compounds representing different chemical classes and one wastewater sample in the course of a 10day uninterrupted flow of tap water was demonstrated (Fig. 9.4b).

A similar biosensor concept was described by Charrier et al., wherein the bioluminescent reporter bacteria were immobilized in an agarose gel [77, 78]. The device comprised four main sections: (a) a transparent polycarbonate removable card containing agarose-immobilized bacteria in 64 wells, (b) a Peltier block on which

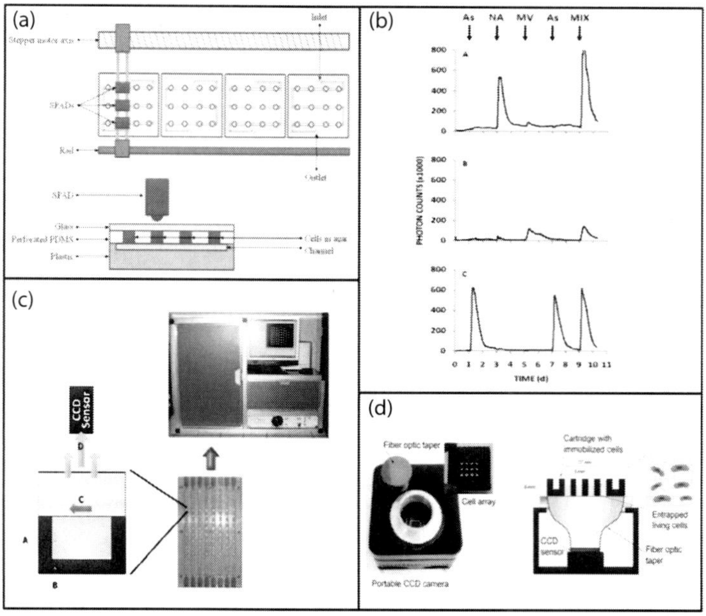

Figure 9.4 (a) Partial scheme of the biosensor device reported by Elad et al. [58]. Top: a top view of the flow-through chambers, the single-photon avalanche diode (SPAD) detectors, and the stepper axis (upper panel); bottom: a cross section of a flow-through chamber and a SPAD detector. Each flow-through chamber is constructed of three layers (glass, PDMS, and poly[methyl methacrylate]), which form a flow channel with 12 wells in its path. (b) Detection of model toxicants in a 10-day monitoring experiment. Signals emitted by (A) *recA*, (B) *micF*, and (C) *arsR* reporters in the course of a 10-day tap water flow are shown. Arrows mark a 2 h pulse of tap water spiked with nalidixic acid (NA), paraquat (MV), As(III), or a mixture of all three (MIX). Curves represent averages of 12 replicate wells smoothed by a simple moving median filter. Panels a and b are reprinted by permission from the American Chemical Society (Copyright 2011). (c) Photo and schematic representation of the multiwell card designed for the biosensor described by Charrier et al. [77, 78]. A: agarose matrix containing immobilized bacteria; B: polycarbonate macrochip (immobilization support); C: liquid circulation; D: bioluminescence measured with a CCD sensor. Material kindly provided by Grald Thouand. (d) Biosensor device reported by Roda et al. [79]. Left: a cooled CCD camera, the cartridge with immobilized cells and the fiber optic taper; right: schematic view of the clear-bottom cartridge with immobilized cells in contact with the CCD sensor through the fiber optic taper. Reproduced by permission from Elsevier.

the card was incubated and a cooled charge-coupled device (CCD) camera above it, (c) a fluidic network that drives the tested sample to the reporter strains, and (d) a dedicated computer interface that manages all the elements of the biosensor, allowing it to operate autonomously (Fig. 9.4c). The online detection of two different metals, separately and combined, was demonstrated immediately after cell immobilization.

Recently, bioluminescent reporter bacteria were immobilized in a matrix composed of three polymers (1% agarose, 2% poly[vinyl pyrrolidone] [PVP], and 0.1% collagen) in a biosensor device [79], composed of a CCD sensor and a disposable cartridge harboring the bacteria immobilized in an array of 3 × 4 wells, connected via a round fiber optic taper (Fig. 9.4d). The device was capable of multiplex detection of two bioluminescence emission wavelengths emitted by reporter bacteria carrying different bioluminescence reporter genes.

Ahn et al. ingeniously overcame the requirement for predetermination of the deposition pattern [80] by forming a biochip harboring a random array of alginate microbeads of five different bioluminescent reporter bacteria. The microbeads were generated by the electrospraying technique [81] and each microbead encapsulated both a bioluminescent stress-specific reporter bacterial strain and fluorescent microspheres as a self-identification code. In this configuration there is no need for prearrangement of the reporter strains since a single bead can be identified among the randomly distributed beads from the color of its self-identifying fluorescent code.

9.4 Long-Term Maintenance of Viability and Activity

The shelf-life of cell-based biosensors is mainly determined by the viability of the cells and by the degree of preservation of their activity. Viability and activity are clearly related, but viable cells are not necessarily active [82]; understanding this connection may thus be helpful [83, 84]. The main considerations in cell preservation include the cells' history (growth phase, availability of compatible solutes), preservation procedure, storage conditions and revival protocol. A variety of methods has been suggested and developed for enhancing bacterial viability and activity maintenance, some of which are discussed in this section.

Hydrogel encapsulation is an option for limited time storage, usually up to a month at 4°C, since the cells are still metabolically active [79, 80]. For long-term preservation, arresting cell metabolism is preferable, for example by freeze-drying (lyophilization). In this process, the cells are frozen and simultaneously dehydrated by reducing the pressure, causing sublimation of the frozen water. Using this technique, bacteria may be preserved for at least two decades [85]. Protocols for freeze-drying bioluminescent bacteria were already described in the 1980s, such as the use of a cryoprotective medium containing 1% NaCl, 15% lactose, and 2% soluble starch [86]. A biosensor constructed using different freeze-dried bioluminescent reporter bacteria was also described [87]. An important note in that work was that the use of surface-active compounds such as Tween 80 may enhance viability, but it may also negatively affect the reporter's signal. More recently, an optimization of the freeze-drying process and the conditions for the subsequent use of freeze-dried bioluminescent bacteria was reported [88], and freeze-dried bioluminescent reporter bacteria were used for the specific detection of methylmercury [89]. The use of osmoprotectants may enhance the survival of dehydrated bacteria for longer time periods. As already described in Section 9.3.1 above, the use of internal and external osmoprotectants may improve viability and activity maintenance [48]. A different approach for growth-arrest preservation of bacteria is their dehydration in a solution containing latex. During the dehydration process pores are formed in the latex coating surrounding the bacteria, and adhesion to the substrate occurs; bacterial activity after 1year storage at −80°C has been reported, but the molecular mechanism of cell preservation by latex coatings is yet to be deciphered [45, 90, 91].

Another option for vacuum preservation of bacteria, vacuum drying, is similar to freeze-drying but is carried out at ambient temperature [92]. Kuppardt et al. [93] compared vacuum drying and hydrogel encapsulation for reporter bacteria preservation for 12 weeks and concluded that vacuum-dried bacteria in the presence of 34% trehalose and 1.5% PVP demonstrated superior activity.

A preservation approach that is unique only for gram-positive spore formers [94, 95] is characterized by the advantages endowed by the intrinsic resistance of spores, which make them ideal for long-term preservation. It has been shown that spores of *Bacillus*-based reporter bacteria can retain their activity even after 1-year storage in diverse extreme conditions [96].

Additional information on the preservation methods of reporter bacteria can be found elsewhere [97, 98].

9.5 Summary

Genetically engineered bacteria for sensing and reporting on the chemical nature of their surroundings were first reported over 30 years ago [99]; recent years may have witnessed more sophisticated molecular manipulations in strain construction but with only few significant modifications of the basic principle of fusing sensor and reporter elements. In contrast, over the last decade significant breakthroughs were reported in the integration of bioluminescent reporter bacteria with electronic elements for the construction of whole-cell biochips and biosensors. For turning such combinations into usable devices, however, more progress is called for in the development of suitable deposition and immobilization techniques of live cells onto hardware platforms, either on traditional surfaces such as glass and silicon or onto polymerbased substrates. Developing techniques such as soft lithography, as well asbiotic and abiotic surface modifications, will clearly continue to contribute to a successful integration of the cells. Even more challenging will be further studies of both fundamental and applied aspects of long-term viability and activity preservation. Immediate future activity is also foreseen in the development of live cell arrays, marrying whole-cell biosensing with the bioassay revolution led by the development of DNA and protein array chips [100]. Such developments will enhance the applicability of whole-cell biosensors in environmental monitoring, homeland security, highthroughput drug and chemical screening, and medical diagnostics.

Acknowledgments

Research was supported by infrastructure project number 3-4187 of the Ministry of Science and Technology, Israel (MOST), by a Korean

National Research Foundation grant to SB and Man Bock Gu, Korea University, and by the EU Seventh Framework project Biomonar.

References

1. van der Meer, J. R., Belkin, S. (2010). Where microbiology meets microengineering: design and applications of reporter bacteria *Nat. Rev. Micro.*, **8**(7), 511–522.
2. Magrisso, S., Erel, Y., Belkin, S. (2008). Microbial reporters of metal bioavailability. *Microb. Biotechnol.*, **1**, 320–330.
3. Melamed, S., Lalush, C., Elad, T., Yagur-Kroll, S., Belkin, S., Pedahzur, R. (2012). A bacterial reporter panel for the detection and classification of antibiotic substances. *Microb. Biotechnol.*, **5**(4), 536–548.
4. Van Dyk, T. K. (2012). Bioluminescent reporter genes for promoter discovery. In *Microbial Metabolic Engineering* Cheng, Q. (ed.), Springer New York, USA, pp. 87–92.
5. Belkin, S., Smulski, D. R., Dadon, S., Vollmer, A. C., Van Dyk, T. K., Larossa, R. A. (1997). A panel of stress-responsive luminous bacteria for the detection of selected classes of toxicants. *Water Res.*, **31**(12), 3009–3016.
6. Yagur-Kroll, S., Bilic, B., Belkin, S. (2009). Strategies for enhancing bioluminescent bacterial sensor performance by promoter region manipulation. *Microb. Biotechnol.*, **3**(3), 300–310.
7. Nivens, D. E., McKnight, T. E., Moser, S. A., Osbourn, S. J., Simpson, M. L., Sayler, G. S. (2004). Bioluminescent bioreporter integrated circuits: potentially small, rugged and inexpensive whole-cell biosensors for remote environmental monitoring. *J. Appl. Microbiol.*, **96**(1), 33–46.
8. Kintzios, S. E. (2006). Cell-based biosensors in proteomic analysis. *Front. Drug Des. Discovery*, **2**(1), 225–239.
9. Xiaorong, X., Lidstrom, M. E., Parviz, B. A. (2007). Microorganisms for MEMS. *J. Microelectromech. Syst.*, **16**(2), 429–444.
10. Ingham, C. J., van HylckamaVlieg, J. E. (2008). MEMS and the microbe. *Lab Chip*, **8**(10), 1604–1616.
11. Ben-Yoav, H., Elad, T., Shlomovits, O., Belkin, S., Shacham-Diamand, Y. (2009). Optical modeling of bioluminescence in whole cell biosensors. *Biosens. Bioelectron.*, **24**(7), 1969–1973.
12. Banerjee, P., Bhunia, A. K. (2010). Cell-based biosensor for rapid screening of pathogens and toxins. *Biosens. Bioelectron.*, **26**(1), 99–106.

13. Ben-Yoav, H., Melamed, S., Freeman, A., Shacham-Diamand, Y., Belkin, S. (2011). Whole-cell biochips for bio-sensing: integration of live cells and inanimate surfaces. *Crit. Rev. Biotechnol.*, **31**(4), 337–353.
14. Judy, J.W. (2001). Microelectromechanical systems (MEMS): fabrication, design and applications. *Smart Mater. Struct.*, **10**(6), 1115.
15. Bras, M., Dugas, V., Bessueille, F., Cloarec, J. P., Martin, J. R., Cabrera, M. et al. (2004). Optimisation of a silicon/silicon dioxide substrate for a fluorescence DNA microarray. *Biosens. Bioelectron.*, **20**(4), 797–806.
16. Petersen, K. E. (1982). Silicon as a mechanical material. *Proc. IEEE*, **70**(5), 420–457.
17. Qin, M., Hou, S., Wang, L., Feng, X., Wang, R., Yang, Y. et al. (2007). Two methods for glass surface modification and their application in protein immobilization. *Colloids Surf. B Biointerfaces*, **60**(2), 243–249.
18. Ham, H. O., Chung, W. Y., Park, K. H., Lee, J. H., Park, H. G. (2008). Use of modified glasses for a microarray-based diagnosis of BRCA mutations. *Biochip J.*, **2**(2), 111–115.
19. Seidemann, V., Bütefisch, S., Büttgenbach, S. (2002). Fabrication and investigation of in-plane compliant SU8 structures for MEMS and their application to micro valves and micro grippers. *Sens. Actuators A*, **97–98**, 457–461.
20. Calleja, M., Nordstrom, M., Alvarez, M., Tamayo, J., Lechuga, L. M., Boisen, A. (2005). Highly sensitive polymer-based cantilever-sensors for DNA detection. *Ultramicroscopy*, **105**(1–4), 215–222.
21. Brandenburg, A., Curdt, F., Nestler, J., Otto, T., Wunderlich, K., Michel, D. (2009). Integrated point of care testing system based on low cost polymer biochips. In *Advanced Biomedical and Clinical Diagnostic Systems VII* SPIE San Jose, CA, USA, pp. 716902–716911.
22. Shacham-Diamand, Y., Krylov, S., Shmilovich, T., Almog, R. O., Fishelson, N., Sverdlov, Y. et al. (2009). Metallization technologies and strategies for plastic based biochips, sensors and actuators for healthcare and medical applications. *ECS Trans.*, **23**(1), 243–254.
23. Yang, C. H., Kuo, L. S., Chen, P. H., Yang, C. R., Tsai, Z. M. (2012). Development of a multilayered polymeric DNA biosensor using radio frequency technology with gold and magnetic nanoparticles. *Biosens. Bioelectron.*, **31**(1), 349–356.
24. Makamba, H., Kim, J. H., Lim, K., Park, N., Hahn, J. H. (2003). Surface modification of poly(dimethylsiloxane) microchannels. *Electrophoresis*, **24**(21), 3607–3619.

25. Zhou, J., Ellis, A. V., Voelcker, N. H. (2010). Recent developments in PDMS surface modification for microfluidic devices. *Electrophoresis*, **31**(1), 2–16.

26. Zhou, J., Khodakov, D. A., Ellis, A. V., Voelcker, N. H. (2012). Surface modification for PDMS-based microfluidic devices. *Electrophoresis*, **33**(1), 89–104.

27. Xia, Y., Whitesides, G. M. (1998). Soft lithography. *Annu. Rev. Mater. Res.*, **28**(1), 153–184.

28. Kane, R. S., Takayama, S., Ostuni, E., Ingber, D. E., Whitesides, G. M. (1999). Patterning proteins and cells using soft lithography. *Biomaterials*, **20**(23–24), 2363–2376.

29. Whitesides, G. M., Ostuni, E., Takayama, S., Jiang, X., Ingber, D. E. (2001). Soft lithography in biology and biochemistry. *Annu. Rev. Biomed. Eng.*, **3**(1), 335–373.

30. Qin, D., Xia, Y., Whitesides, G. M. (2010). Soft lithography for micro- and nanoscale patterning. *Nat. Protoc.*, **5**(3), 491–502.

31. Moreau, W. M. (1988). *Semiconductor Lithography: Principles, Practices, and Materials*, Plenum Press, New York, USA.

32. Kumar, A., Whitesides, G. M. (1993). Features of gold having micrometer to centimeter dimensions can be formed through a combination of stamping with an elastomeric stamp and an alkanethiol ink followed by chemical etching. *Appl. Phys. Lett.*, **63**, 2002–2004.

33. Xia, Y. N., Kim, E., Zhao, X. M., Rogers, J. A., Prentiss, M., Whitesides, G. M. (1996). Complex optical surfaces formed by replica molding against elastomeric masters. *Science*, **273**(5273), 347–349.

34. Zhao, X. M., Xia, Y., Whitesides, G. M. (1996). Fabrication of three-dimensional microstructures: microtransfer molding. *Adv. Mater.*, **8**(10), 837–840.

35. Kim, E., Xia, Y., Whitesides, G. M. (1995). Polymeric microstructures formed by moulding in capillaries. *Nature*, **376**, 581–584.

36. Kim, E., Xia, Y., Zhao, X. M., Whitesides, G. M. (1997). Solvent-assisted microcontact molding: a convenient method for fabricating three-dimensional structures of polymeric materials. *Adv. Mater.*, **9**, 651–654.

37. Rogers, J. A., Paul, K. E., Jackman, R. J., Whitesides, G. M. (1997). Using an elastomeric phase mask for sub-100 nm photolithography in the optical near field. *Appl. Phys. Lett.*, **70**, 2658–2660.

38. Zaumseil, J., Meitl, M. A., Hsu, J. W. P., Acharya, B. R., Baldwin, K. W., Loo, Y. L., Rogers, J. A. (2003). Three-dimensional and multilayer nanostructures formed by nanotransfer printing. *Nano Lett.*, **3**(9), 1223–1227.
39. Jeon, S., Menard, E., Park, J. U., Maria, J., Meitl, M., Zaumseil, J., Rogers, J. A. (2004). Three-dimensional nanofabrication with rubber stamps and conformable photomasks. *Adv. Mater.*, **16**(15), 1369–1373.
40. Childs, W. R., Nuzzo, R. G. (2002). Decal transfer microlithography: a new soft-lithographic patterning method. *J. Am. Chem. Soc.*, **124**, 13583–13596.
41. Xu, Q., Rioux, R. M., Dickey, M. D., Whitesides, G. M. (2008). Nanoskiving: a new method to produce arrays of nanostructures. *Acc. Chem. Res.*, **41**, 1566–1577.
42. Lipomi, D. J., Martinez, R. V., Kats, M. A., Kang, S. H., Kim, P., Aizenberg, J. et al. (2011). Patterning the tips of optical fibers with metallic nanostructures using nanoskiving. *Nano Lett.*, **11**(2), 632–636.
43. Bradley, R. R., Christina, M. O., Jason, A. B., Daniel, Y., Barry, J. S. (2006). Jet-based methods to print living cells. *Biotechnol. J.*, **1**(9), 930–948.
44. Calvert, P. (2007). Materials science: printing cells. *Science*, **318**(5848), 208–209.
45. Flickinger, M. C., Schottel, J. L., Bond, D. R., Aksan, A., Scriven, L. E. (2008). Painting and printing living bacteria: engineering nanoporousbiocatalytic coatings to preserve microbial viability and intensify reactivity. *Biotechnol. Prog.*, **23**(1), 2–17.
46. Ringeisen, B. R., Fitzgerald, L. A., Lizewski, S. E., Biffinger, J. C., Wu, P. K. (2010). Bacterial cell printing. In *Cell and Organ Printing*, Ringeisen, B. R., Spargo, B. J., Wu, P. K. (eds.), Springer Netherlands, pp. 243–256.
47. Merrin, J., Leibler, S., Chuang, J. S. (2007). Printing multistrain bacterial patterns with a piezoelectric inkjet printer. *PLOS ONE*, **2**(7), e663.
48. Melamed, S., Ceriotti, L., Weigel, W., Rossi, F., Colpo, P., Belkin, S. (2011). A printed nanolitre-scale bacterial sensor array. *Lab Chip*, **11**(1), 139–146.
49. Okamoto, T., Suzuki, T., Yamamoto, N. (2000). Microarray fabrication with covalent attachment of DNA using bubble jet technology. *Nat. Biotechnol.*, **18**(4), 438–441.
50. Allain, L. R., Stratis-Cullum, D. N., Vo-Dinh, T. (2004). Investigation of microfabrication of biological sample arrays using piezoelectric and bubble-jet printing technologies. *Anal. Chim. Acta*, **518**(1–2), 77–85.

51. Vidal, O., Longin, R., Prigent-Combaret, C., Dorel, C., Hooreman, M., Lejeune, P. (1998). Isolation of an *Escherichia coli* K-12 mutant strain able to form biofilms on inert surfaces: involvement of a new *ompR* allele that increases curli expression. *J. Bacteriol.*, **180**(9), 2442–2449.

52. Schembri, M. A., Klemm, P. (2001). Biofilm formation in a hydrodynamic environment by novel FimH variants and ramifications for virulence. *Infect. Immun.*, **69**(3), 1322–1328.

53. Disney, M. D., Seeberger, P. H. (2004). The use of carbohydrate microarrays to study carbohydrate-cell interactions and to detect pathogens. *Chem. Biol.*, **11**(12), 1701–1707.

54. Müller, R., Ruhl, S., Hiller, K. A., Schmalz, G., Schweikl, H. (2008). Adhesion of eukaryotic cells and *Staphylococcus aureus* to silicon model surfaces. *J. Biomed. Mater. Res. A*, **84A**(3), 817–827.

55. Nyamjav, D., Rozhok, S., Holz, R. C. (2010). Immobilization of motile bacterial cells via dip-pen nanolithography. *Nanotechnology*, **21**(23), 235105.

56. Fesenko, D. O., Nasedkina, T. V., Mirzabekov, A. D. (2001). A bacterial microchip: the principle of operation as exemplified by detection of antibiotics. *Dokl. Biochem. Biophys.*, **381**(6), 427–429.

57. Fesenko, D. O., Nasedkina, T. V., Prokopenko, D. V., Mirzabekov, A. D. (2005). Biosensing and monitoring of cell populations using the hydrogel bacterial microchip. *Biosens. Bioelectron.*, **20**(9), 1860–1865.

58. Elad, T., Almog, R., Yagur-Kroll, S., Levkov, K., Melamed, S., Shacham-Diamand, Y., Belkin, S. (2011). Online monitoring of water toxicity by use of bioluminescent reporter bacterial biochips. *Environ. Sci. Technol.*, **45**(19), 8536–8544.

59. An, Y. H., Friedman, R. J. (1998). Concise review of mechanisms of bacterial adhesion to biomaterial surfaces. *J. Biomed. Mater. Res.*, **43**(3), 338–348.

60. Choi, C. H., Lee, J. H., Hwang, T. S., Lee, C. S., Kim, Y. G., Yang, Y. H., Huh, K. (2010). Preparation of bacteria microarray using selective patterning of polyelectrolyte multilayer and poly(ethylene glycol)-poly(lactide) diblock copolymer. *Macromol. Res.*, **18**(3), 254–259.

61. Kempf, B., Bremer, E. (1998). Uptake and synthesis of compatible solutes as microbial stress responses to high-osmolality environments. *Arch. Microbiol.*, **170**(5), 319–330.

62. Welsh, D. T., Herbert, R. A. (1999). Osmotically induced intracellular trehalose, but not glycine betaine accumulation promotes desiccation tolerance in *Escherichia coli*. *FEMS Microbiol. Lett.*, **174**(1), 57–63.

63. Manzanera, M., Vilchez, S., Tunnacliffe, A. (2004). High survival and stability rates of *Escherichia coli* dried in hydroxyectoine. *FEMS Microbiol. Lett.*, **233**(2), 347–352.

64. Suo, Z., Avci, R., Yang, X., Pascual, D. W. (2008). Efficient immobilization and patterning of live bacterial cells. *Langmuir*, **24**(8), 4161–4167.

65. Scott, J. R., Zähner, D. (2006). Pili with strong attachments: gram-positive bacteria do it differently. *Mol. Microbiol.*, **62**(2), 320–330.

66. Proft, T., Baker, E. (2009). Pili in gram-negative and gram-positive bacteria: structure, assembly and their role in disease. *Cell. Mol. Life Sci.*, **66**(4), 613–635.

67. Kline, K. A., Dodson, K. W., Caparon, M. G., Hultgren, S. J. (2010). A tale of two pili: assembly and function of pili in bacteria. *Trends Microbiol.*, **18**(5), 224–232.

68. Kang, H. J., Baker, E. N. (2012). Structure and assembly of gram-positive bacterial pili: unique covalent polymers. *Curr. Opin. Struct. Biol.*, **22**(2), 200–207.

69. Pratt, L. A., Kolter, R. (1998). Genetic analysis of *Escherichia coli* biofilm formation: roles of flagella, motility, chemotaxis and type I pili. *Mol. Microbiol.*, **30**(2), 285–293.

70. Saldaña, Z., Xicohtencatl-Cortes, J., Avelino, F., Phillips, A. D., Kaper, J. B., Puente, J. L., Girón, J. A. (2009). Synergistic role of curli and cellulose in cell adherence and biofilm formation of attaching and effacing *Escherichia coli* and identification of Fis as a negative regulator of curli. *Environ. Microbiol.*, **11**(4), 992–1006.

71. Colpo, P., Ruiz, A., Ceriotti, L., Rossi, F., Belkin, S., Gu, M. B. (2010). Surface functionalization for protein and cell patterning. In *Whole Cell Sensing Systems I*, Springer Berlin, Heidelberg, pp. 109–130.

72. Fesenko, D. O., Nasedkina, T. V., Chudinov, A. V., Prokopenko, D. V., Iurasov, R. A., Zasedatelev, A. S. (2005). Alginate gel biochip for real-time monitoring of intracellular processes in bacterial and yeast cells. *Mol. Biol.*, **39**(1), 84–89.

73. Mirsaidov, U., Scrimgeour, J., Timp, W., Beck, K., Mir, M., Matsudaira, P., Timp, G. (2008). Live cell lithography: using optical tweezers to create synthetic tissue. *Lab Chip*, **8**(12), 2174–2181.

74. Desimone, M. F., Alvarez, G. S., Foglia, M. L., Diaz, L. E. (2009). Development of sol-gel hybrid materials for whole cell immobilization. *Recent Pat. Biotechnol.*, **3**, 55–60.

75. Shin, H. (2012). Agarose-gel-immobilized recombinant bacterial biosensors for simple and disposable on-site detection of phenolic compounds. *Appl. Microbiol. Biotechnol.*, **93**(5), 1895–1904.

76. Depagne, C., Roux, C., Coradin, T. (2011). How to design cell-based biosensors using the sol–gel process. *Anal. Bioanal. Chem.*, **400**(4), 965–976.

77. Charrier, T., Chapeau, C., Bendria, L., Picart, P., Daniel, P., Thouand, G. (2011). A multi-channel bioluminescent bacterial biosensor for the on-line detection of metals and toxicity. Part II: technical development and proof of concept of the biosensor. *Anal. Bioanal. Chem.*, **400**(4), 1061–1070.

78. Charrier, T., Durand, M. J., Jouanneau, S., Dion, M., Pernetti, M., Poncelet, D., Thouand, G. (2011). A multi-channel bioluminescent bacterial biosensor for the on-line detection of metals and toxicity. Part I: design and optimization of bioluminescent bacterial strains. *Anal. Bioanal. Chem.*, **400**(4), 1051–1060.

79. Roda, A., Cevenini, L., Michelini, E., Branchini, B. R. (2011). A portable bioluminescence engineered cell-based biosensor for on-site applications. *Biosens. Bioelectron.*, **26**(8), 3647–3653.

80. Ahn, J. M., Kim, J. H., Kim, J. H., Gu, M. B. (2010). Randomly distributed arrays of optically coded functional microbeads for toxicity screening and monitoring. *Lab Chip*, **10**(20), 2695–2701.

81. Halle, J. P., Leblond, F. A., Pariseau, J. F., Jutras, P., Brabant, M. J., Lepage, Y. (1994). Studies on small (<300 microns) microcapsules: II; parameters governing the production of alginate beads by high voltage electrostatic pulses. *Cell Transplant.*, **3**(5), 365–372.

82. Kell, D. B., Kaprelyants, A. S., Weichart, D. H., Harwood, C. R., Barer, M. R. (1998). Viability and activity in readily culturable bacteria: a review and discussion of the practical issues. *Antonie Leeuwenhoek*, **73**(2), 169–187.

83. Hayes, C. S., Low, D. A. (2009). Signals of growth regulation in bacteria. *Curr. Opin. Microbiol.*, **12**(6), 667–673.

84. Trevors, J. T. (2011). Viable but non-culturable (VBNC) bacteria: gene expression in planktonic and biofilm cells. *J. Microbiol. Methods*, **86**(2), 266–273.

85. Miyamoto-Shinohara, Y., Sukenobe, J., Imaizumi, T., Nakahara, T. (2008). Survival of freeze-dried bacteria. *J. Gen. Appl. Microbiol.*, **54**(1), 9–24.

86. Janda, I., Opekarová, M. (1989). Long-term preservation of active luminous bacteria by lyophilization. *J. Biolumin. Chemilumin.*, **3**(1), 27–29.

87. Choi, S. H., Gu, M. B. (2002). A portable toxicity biosensor using freeze-dried recombinant bioluminescent bacteria. *Biosens. Bioelectron.*, **17**(5), 433–440.

88. Camanzi, L., Bolelli, L., Maiolini, E., Girotti, S., Matteuzzi, D. (2010). Optimal conditions for stability of photoemission and freeze drying of two luminescent bacteria for use in a biosensor. *Environ. Toxicol. Chem.*, **30**(4), 801–805.
89. Rantala, A., Utriainen, M., Kaushik, N., Virta, M., Välimaa, A.-L., Karp, M. (2011). Luminescent bacteria-based sensing method for methylmercury specific determination. *Anal. Bioanal. Chem.*, **400**(4), 1041–1049.
90. Schottel, J., Orwin, P., Anderson, C., Flickinger, M. (2008). Spatial expression of a mercury-inducible green fluorescent protein within a nanoporous latex-based biosensor coating. *J. Ind. Microbiol. Biotechnol.*, **35**(4), 283–290.
91. Gosse, J. L., Flickinger, M. C. (2011). Uniform lab-scale biocatalytic-nanoporous latex coatings for reactive microorganisms. In *Nanoscale Biocatalysis*. Wang, P. (ed.), Humana Press, New York, USA, pp. 213–222.
92. Stocker, J., Balluch, D., Gsell, M., Harms, H., Feliciano, J., Daunert, S. et al. (2003). Development of a set of simple bacterial biosensors for quantitative and rapid measurements of arsenite and arsenate in potable water. *Environ. Sci. Technol.*, **37**(20), 4743–4750.
93. Kuppardt, A., Chatzinotas, A., Breuer, U., van der Meer, J., Harms, H. (2009). Optimization of preservation conditions of As (III) bioreporter bacteria. *Appl. Microbiol. Biotechnol.*, **82**(4), 785–792.
94. Date, A., Pasini, P., Daunert, S. (2007). Construction of spores for portable bacterial whole-cell biosensing systems. *Anal. Chem.*, **79**(24), 9391–9397.
95. Date, A., Pasini, P., Daunert, S. (2010). Fluorescent and bioluminescent cell-based sensors: strategies for their preservation. *Adv. Biochem. Eng. Biotechnol.*, **117**, 57–75.
96. Sangal, A., Pasini, P., Daunert, S. (2011). Stability of spore-based biosensing systems under extreme conditions. *Sens. Actuators B*, **158**(1), 377–382.
97. Bjerketorp, J., Hakansson, S., Belkin, S., Jansson, J. K. (2006). Advances in preservation methods: keeping biosensor microorganisms alive and active. *Curr. Opin. Biotechnol.*, **17**(1), 43–49.
98. Michelini, E., Roda, A. (2012). Staying alive: new perspectives on cell immobilization for biosensing purposes. *Anal. Bioanal. Chem.*, **402**(5), 1785–1797.

99. Quillardet, P., Huisman, O., D'Ari, R., Hofnung, M. (1982). SOS chromotest, a direct assay of induction of an SOS function in *Escherichia coli* K-12 to measure genotoxicity. *Proc. Natl. Acad. Sci. U S A*, **79**(19), 5971–5975.

100. Melamed, S., Elad, T., Belkin, S. (2012). Microbial sensor cell arrays. *Curr. Opin. Biotechnol.*, **23**(1), 2–8.

Chapter 10

Luminescent Biosensing Systems Based on Genetically Engineered Spore-Forming Bacteria

Leslie D. Knecht,[a] Patrizia Pasini,[b] and Sylvia Daunert[b]

[a]*Department of Chemistry, University of Miami, Miami, FL 33146, USA*
[b]*Miller School of Medicine, Department of Biochemistry and Molecular Biology, University of Miami, Miami FL 33136, USA*
sdaunert@med.miami.edu

10.1 Introduction

When bacteria, such as certain species of *Bacillus* and *Clostridium*, are exposed to extreme environmental conditions, including lack of nutrients, they are able to form spores that can survive for long periods of time, with some reports of viable spores found that were 25 million years old [1]. *Clostridium* species are gram-positive, obligatory anaerobic bacteria. For this reason, clostridia spores do not germinate until they are in a hypoxic environment, making them ideal vehicles for therapeutics at hypoxic tumor sites [2]; however, due to their anaerobic nature, *Clostridium* species have limited uses in bioanalytical applications [3]. On the other end, *Bacillus* species

Bioluminescent Microbial Biosensors: Design, Construction, and Implementation
Edited by Gérald Thouand and Robert S. Marks

ISBN 978-981-4613-65-1 (Hardcover), 978-981-4613-66-8 (eBook)
www.panstanford.com

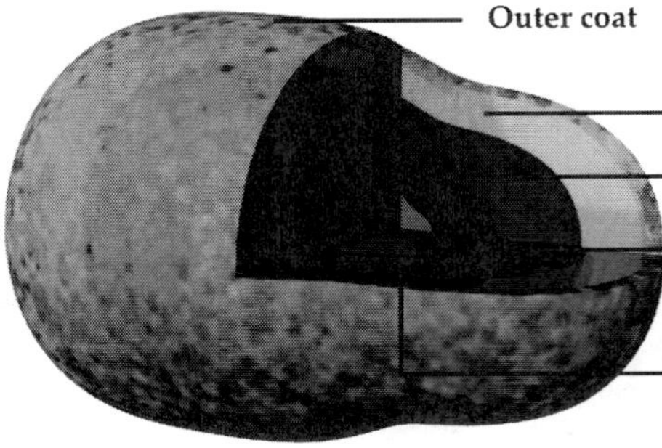

Figure 10.1 Schematic showing the components of a spore.

have been better characterized and have been found to be more amenable for analytical applications and are thus the emphasis of this chapter.

At the onset of sporulation in *Bacillus* species, the cell produces a prespore and a mother cell by undergoing an asymmetrical division. The prespore is then engulfed by the mother cell, where it is dehydrated, mineralized, and coated with protective layers (Fig. 10.1) [4]. The exterior of the spore, or the spore coat, comprises multilayered proteins. The inner and outer coats protect the spore from various environmental stresses, including organic solvents and lysozymes [5]. It has been shown that the coat proteins provide flexibility to the spore and allow for the spore to expand and contract, accommodating for dehydration when the spore is formed and rehydration when germination occurs [6]. The interior of the spore consists of a thick layer of peptidoglycans that surround a forespore, which is the part of the mother cell that commits to sporulation. This thick layer, called the spore cortex, provides a template for peptidoglycan synthesis upon germination and helps to prevent the loss of cellular integrity after germination [7]. Additionally, the cortex acts as a desiccator to keep the interior DNA dry during sporulation [6]. The core of the spore is comprised of α/β small acid soluble proteins (SASPs) complexed to the DNA and protects the spore DNA from ultraviolet (UV) light [8] and extreme temperatures [8, 9]. These SASPs are double-stranded DNA-binding proteins that, upon binding, force the DNA into an A-like conformation.

The binding has been shown to protect the DNA from enzymatic and chemical cleavage and reduce the amount of depurination [10]. Overall, these layers protect the valuable DNA inside the spore from environmental damage due to UV light, extreme temperatures, irradiation, toxic compounds, extreme humidity levels, and other harmful factors. Additionally, the spores of some *Bacillus* species have a further protective outer layer composed of proteins, lipids, and carbohydrates, called an exosporium. In bacterial species that are human or animal pathogens, this layer protects their spores from the host immune response.

Germination of the spore occurs when receptors on the spore surface sense nutrients in the surrounding environment. These germination receptors respond to specific nutrients and trigger the onset of germination. *Bacillus subtilis* spores have been found to germinate in the presence of L-alanine, L-valine, or a combination of L-asparagine, D-glucose, D-fructose, and K^+ [11]. The detailed knowledge of the spore at the functional, structural, and molecular levels, as well as the robustness of spores, makes them ideal candidates for use as vehicles for whole-cell biosensing systems and as platforms for display of recombinant heterologous proteins for applications in bioanalysis.

10.2 Whole-Cell Biosensors

Inducible bacterial whole-cell biosensors are comprised of genetically engineered bacterial cells harboring a plasmid that bears a reporter element under the control of a specific operator/promoter (O/P). It is necessary to also have a regulatory/recognition element coding for a protein that will bind to an analyte of interest and, thus, either bind to (positive regulation) or release (negative regulation) the O/P region to begin transcription of the reporter element. Within a certain concentration range, this interaction results in an analyte dose-dependent response that is detectable by a color change, light emission, or other quantifiable signal, depending on the nature of the reporter and detection method employed (Fig. 10.2). Noninducible bacterial whole-cell biosensors rely on the toxicity of the molecule or sample of interest. These either naturally

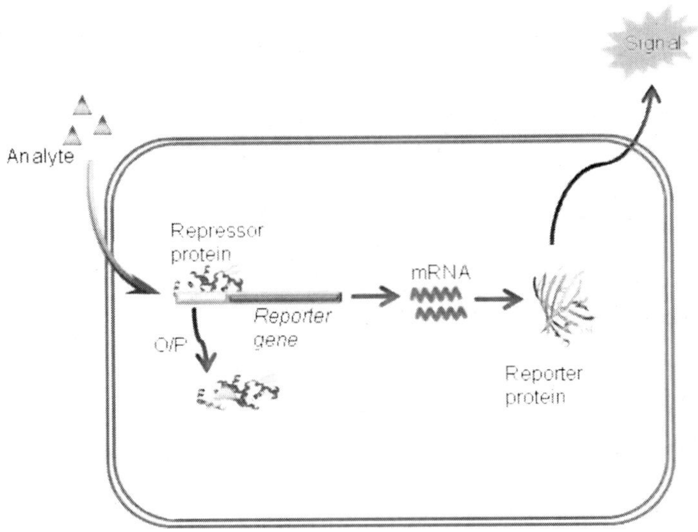

Figure 10.2 Inducible whole-cell biosensors harbor plasmids carrying reporter elements under the control of an O/P region. Upon analyte binding to a recognition/regulatory protein, the protein either releases the O/P region (negative regulation), as shown in the schematic, or binds to the O/P region (positive regulation) to begin transcription of downstream reporter genes. The result is a dose-dependent expression of the reporter protein, accompanied by a dose-dependent signal production based on analyte concentration.

produce a signal, usually bioluminescence, or contain a plasmid with a reporter protein that is constitutively expressed. When a toxic compound that is lethal to the cells is introduced to the system, the cells die, which results in a dose-dependent decrease in signal of the reporter protein. Inducible whole-cell biosensors have been used for detection of a variety of analytes [12–14]. These biosensing systems have proven to be valuable analytical tools due to their selectivity, sensitivity, rapidity, ease of use, and cost-effectiveness; however, there are still obstacles, especially for field analysis, due to a lack of stability of the sensing cells for long-term storage under various environmental conditions. To overcome these issues, spores have been employed as vehicles for preservation, storage, transport, and onsite application of whole-cell biosensors.

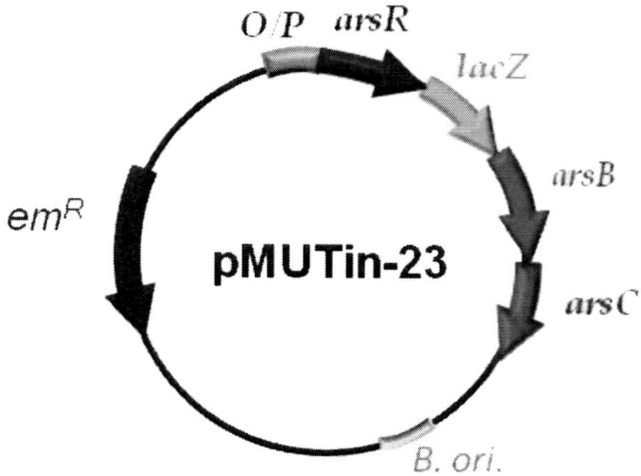

Figure 10.3 Plasmid pMUTin-23. Upon binding of As(III), the negative transcriptional regulatory protein ArsR releases the O/P region to begin transcription of the genes *lacZ*, which encodes for β-galactosidase, and *arsB* and *arsC*, which help confer resistance to arsenic. Adapted with permission from Ref. [16]. Copyright (2007) American Chemical Society.

10.2.1 *Arsenic Spore-Based Whole-Cell Biosensor*

Arsenic is an environmental contaminant that leads to cancer, vascular diseases, and dermatitis in humans [15]. Date et al. developed a whole-cell biosensor encapsulated in a spore for arsenic detection and evaluated the performance of the spore-based sensing system over multiple germination/sporulation cycles, as well as after several months of storage [16]. To do this, *B. subtilis* cells were transformed with the plasmid pMUTin-23 (Fig. 10.3) [17]. *B. subtilis* is one of the well-studied spore-forming bacteria. Additionally, these cells have a natural competency at a specific optical density at 600 nm (OD_{600}), thus requiring no extra equipment, such as an electroporator, or special preparation, which is usually required to make chemically competent cells for plasmid transformation. The plasmid pMUTin-23 consists of three genes that confer resistance to arsenic, *arsR*, *arsB*, and *arsC*, as well as the *lacZ* gene encoding for the reporter protein β-galactosidase, which is under control of the ArsR regulatory protein and the *ars* operon O/P [18]. ArsR is a regulatory repressor protein that controls the expression of the *ars* operon. When no arsenic (III) is present, ArsR is bound to the O/P region of the DNA and prevents the transcription of the *ars* operon; however, when arsenic (III) is present, it binds to ArsR, which is then released from the DNA and transcription occurs. ArsB is a membrane protein that pumps Ar(III) out of the cell using an electrochemical gradient and ArsC is a cytoplasmic protein that reduces As(V) to As(III).

The *B. subtilis* cells harboring the plasmid pMUTin-23 were sporulated using the exhaustion method [19]. The cells were suspended in Luria-Bertani (LB) broth and incubated at 37°C with shaking at 250 rpm for 4 days. During this time, the cells grew and exhausted the nutrient supply of the LB broth, which initiated sporulation. After 4 days, the spores were harvested by centrifugation at 10,000 g for 10 min at 4°C, washed twice with deionized distilled water, and resuspended in deionized distilled water. The spores were then sonicated for 5 min with a 550 sonic dismembrator from Fisher Scientific (Pittsburg, PA) to separate the endospore from the mother cell membrane. The sonicated mixture was again centrifuged at 10,000 g for 10 min at 4°C and decanted to remove the cell debris from the spores. The spores were then resuspended in deionized distilled water to an OD_{600} of 0.8 and stored at room temperature. To confirm the morphology and size of the spores, scanning electron microscopy (SEM) images of the spores were acquired (Fig. 10.4).

Assays were then performed to determine the analytical performance of the sporulated biosensor by evaluating the dynamic range and the detection limit before sporulation and at various time points after sporulation. To this end, a fresh colony of the *B. subtilis* ars-23 strain or 20 μL of the spore suspension were used to inoculate 300 mL of fresh LB broth containing 50 μg/mL erythromycin. The inoculated cultures were grown overnight at 37°C, 250 rpm until the OD_{600} reached a value of 0.6–0.7. Then, 1 mL aliquots of the cells from the colony or the germinated cells from the spore suspension were transferred to culture tubes. To test the response of the whole-cell biosensor, standard solutions ranging from 1×10^{-4}–1×10^{-9} M of potassium antimonyl tartrate, sodium arsenate, and sodium arsenite were prepared in deionized distilled water from a 1×10^{-2} M deionized distilled water stock

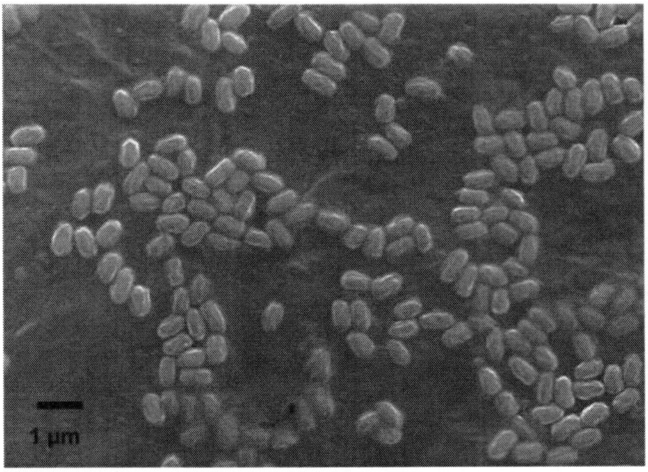

Figure 10.4 Scanning electron microscopy (SEM) of *B. subtilis* spores. Adapted with permission from Ref. [16]. Copyright (2007) American Chemical Society.

solution of each. Triplicate samples were made by adding three 200 µL aliquots of each standard solution to the culture tubes containing the cells. Blank samples were created by adding three 200 µL aliquots of deionized distilled water to cells. The samples were then incubated at 37°C, 250 rpm for 1 h. After incubation, the cells were centrifuged at 8000 g for 5 min at 4°C. The resulting pelleted cells were washed in 100 mM phosphate buffer, pH 7.4, resuspended in the same buffer containing 0.2% (w/v) triton X-100, and placed in a shaker at 37°C, 250 rpm for 1 h. This incubation step was to allow for β-galactosidase to be released into the media after cell lysis. After incubation, 30 µL of the cell lysate was added in triplicate to 100 µL of the chemiluminescent substrate for β-galactosidase in a 96-well microtiter plate and incubated at 37°C, 250 rpm for 1 h. To trigger the chemiluminescent signal, 100 µL of the accelerator-II solution provided with the chemiluminescent β-galactosidase assay kit was injected into the substrate–enzyme mixture, and after a 20 s delay that allowed for enough time to reach maximum signal intensity, the signal was measured over a time period of 3 s using a PolarStar Optima microtiter plate reader from BMG Labtech (Durham, NC). This microtiter plate reader has the ability to perform fluorescence, bioluminescence, chemiluminescence, and absorbance measurements. It employs a photomultiplier tube (PMT) as the detector and has a built-in automated injection system. PMTs comprise dynodes with increasing positive charges. When the light (photons) hits the first dynode, it produces electrons. These electrons are drawn toward the next, more positively charged dynode, which generates even more electrons. This cascade continues, thus allowing for the signal to be amplified. Due to this amplification, the PMT is a valuable, sensitive tool for light detection. The automated injection system is very important when performing experiments, such as the one described above, where a specific time delay is needed between the addition of a reagent and the reading of the signal. The chemiluminescence detection method has an advantage over fluorescence detection in that it provides a high signal-to-noise ratio. This method, however, also demonstrates the disadvantage of using β-galactosidase as the reporter due to the need to add an exogenous substrate and to efficiently lyse the cells to detect the signal. Additionally, the ability to monitor the production of proteins in vivo in real time is lacking. Considerations, such as the ones mentioned above, should be made when determining a proper reporter protein for a whole-cell sensing system.

The results of the experiments showed that the whole-cell biosensor was able to respond to sodium arsenite, potassium antimonyl tartrate, and, due to the ArsC protein that reduces arsenate to arsenite, sodium arsenate. Both sodium arsenite and arsenate were detectable down to 1×10^{-7} M, while potassium antimonyl tartrate was detectable down to 1×10^{-8} M. Concentrations of the analyte above 1×10^{-4} M showed a decrease in the signal, indicating that this concentration was toxic to the cells. Analyte toxicity must always be taken into account when developing whole-cell-based assays. The detection limits before sporulation were 2.8×10^{-8}, 1.6×10^{-7}, and 2.8×10^{-7} M for antimonite, arsenate, and arsenite, respectively. Inter- and intra-assay reproducibility was tested, resulting in a coefficient of variation (CV) of 4%–14%. The interassay reproducibility demonstrates how close values are between multiple assays performed at different times, while the intra-assay reproducibility demonstrates how close values are between replicates in a specific assay at a particular time. These values are important when developing an analytical method, especially with spores that can be germinated and sporulated, to indicate the viability of exploiting the spore as a vehicle for biosensors. Additionally, the selectivity of the sensor was evaluated. The aforementioned protocol was carried out, as previously described, except that various concentrations of sodium sulfate, potassium ferricyanide, potassium dichromate, and sodium arsenite (which is the reference compound) were added to the cultures. These compounds were chosen for their size, charge, and structure similarity to antimonite and arsenite; all of them showed a negligible response when compared to sodium arsenite, thus confirming the selectivity of the sensor.

Importantly, the sensing cells maintained their analytical features after sporulation and subsequent germination. This observation prompted the authors to evaluate the validity of exposing the bacteria to several sporulation–germination cycles in 1-month intervals. Indeed, the analytical performance did not vary between the sporulation–germination cycles. Furthermore, the analytical characteristics such as the limits of detection, dynamic ranges, and CVs did not change significantly as compared to the data obtained before sporulation, even after 2 years of storage at room temperature [20]. These data support the use of spores for long-term storage of sensing bacterial cells.

10.2.2 *Zinc Spore-Based Whole-Cell Biosensor*

Date et al. also developed a whole-cell biosensor for detection of the essential element zinc using spores from *Bacillus megaterium* as the vehicle [16]. A deficiency in the amount of zinc present in the human body can result in impaired wound healing, decreased growth rate and development in infants, and reproductive failure, among others [21]. To prepare the sensor, the plasmid pSD202 was created by inserting the genes for *egfp* and *smtB* into restriction sites *KpnI-SphI* of the plasmid pMM1522 (Fig. 10.5). The *egfp* gene, which encodes for the enhanced green fluorescent protein (EGFP), was amplified from the plasmid pEGFP, while the *smt* operon O/P

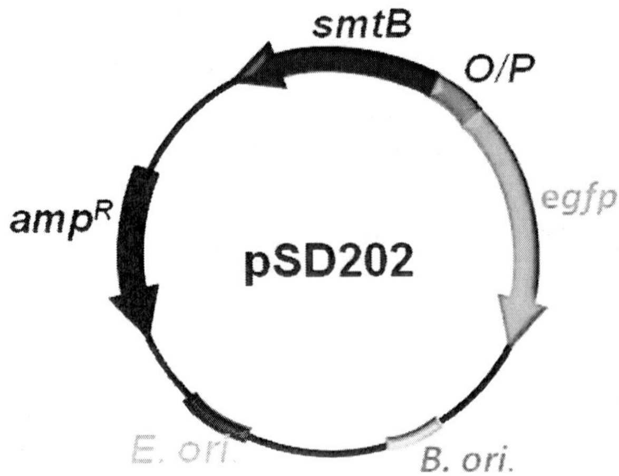

Figure 10.5 Plasmid pSD202. The *egfp* gene is under the control of the SmtB protein and the O/P region. SmtB binds zinc and releases the O/P to begin transcription of EGFP. Adapted with permission from Ref. [16]. Copyright (2007) American Chemical Society.

and *smtB* gene, which encodes for the SmtB zinc-binding protein that regulates the *smt* operon of *Synechococcus* PCC7942, were isolated from the plasmid pJLE3C using polymerase chain reaction (PCR) [17]. SmtB is a repressor protein that, upon binding to zinc, releases the O/P region to begin the transcription of *smtA*. In this construct, however, the *smtA* gene was replaced by the *egfp* gene. Therefore, the expression of the EGFP reporter protein occurs in a dose-dependent manner upon zinc being introduced to the system. After gene construction, the plasmid was transformed into the *B. megaterium* strain WH320 (Mobitec, Boca Rotan, FL). *B. megaterium* is an ideal candidate for transformation of larger plasmids due to its large size and the fact that it does not produce proteases to degrade foreign proteins [22]. Although the strain is developed for efficient transformation of larger plasmids, intact *B. megaterium* cells cannot easily be transformed; hence, they are not typically employed as such. Instead, *B. megaterium* protoplasts, where the cell wall has been completely removed, are used. Protoplasts are time consuming and cumbersome to prepare; on the other hand, the commercially available ones, for example, those from Mobitec, are expensive and have a short shelf life of 2 months at −80°C.

The transformed *B. megaterium* cells were sporulated using the same exhaustion method as for the *B. subtilis* cells described above. Again, to test the validity of using spores as vehicles for whole-cell biosensors, the prepared spores or a colony of vegetative cells on an agar plate, never exposed to sporulation/germination, were used to inoculate 5 mL LB broth containing tetracycline (20 μg/mL) and cultured overnight at 37°C and 250 rpm. When the OD_{600} reached 0.7–0.8, each small culture was inoculated into 100 mL fresh LB broth (tetracycline, 20 μg/mL) in a 500 mL flask and grown at 37°C, 250 rpm until the OD_{600} reached 0.8. The cells were then centrifuged at 6000 rpm for 5 min at 4°C. After discarding the supernatant, the pellet was washed twice with M9 minimal salt medium. The pelleted cells were resuspended in M9 medium supplemented with $CaCl_2$, $MgSO_4$, and 10% glucose to obtain an OD_{600} of 0.8, and then 1.8 mL aliquots of cell culture were added to culture tubes. Standard solutions of zinc ranging from 1×10^{-4} to 1×10^{-7} M were made in distilled deionized water by serial dilution starting with a 1×10^{-2} M zinc chloride stock solution in water. The standards and blank (distilled deionized water) were added at a volume of 200 μL in triplicate to the tubes containing the cells and incubated for 1 h at 37°C, 250 rpm. The cells were then harvested by centrifugation at 6000 rpm for 5 min at 4°C; the supernatant was discarded, and the pellet was washed twice with 100 mM phosphate buffer, pH 7.5. The cells were resuspended in the same buffer to a volume of 500 μL and then diluted 1:10 with the same buffer. Next, 200 μL of the diluted cell suspensions was transferred to a 96-well microtiter plate in triplicate and the fluorescent readings were taken on the PMT-based PolarStar Optima luminometer with an excitation wavelength of 480 nm and an emission wavelength of 510 nm. The use of EGFP for detection in this system allows one to overcome the shortcomings of the reporter protein (β-galactosidase) detected by chemiluminescence, which was described in the previous section, because the cells do not have to be lysed to release the reporter and there is no need for the addition of a substrate. Fluorescence detection may suffer from a high background, which can lower the sensitivity of a method. This is due to the presence of fluorescent molecules, such as aromatic amino acids, in the test samples. For this reason, the detection was performed in a buffer instead of LB broth, whose components can create a high fluorescent background signal.

The results of these experiments showed that the whole-cell biosensor had a detection limit of 1×10^{-6} M and a dynamic range of 1×10^{-4}–1×10^{-6} M. Furthermore, these analytical parameters were maintained after 8 months of storage at room temperature in spore form, confirming the successful use of spores as vehicles for whole-cell biosensors. The spores were also subjected to multiple sporulation/germination cycles in 1-month intervals. It was found that the whole-cell biosensor retained the same detection limit and dynamic range of the cells presporulation. The SmtB protein has been previously shown to respond to both Cd^{2+} and Co^{2+} [23]; therefore, the chloride salts of these metals, along with nickel chloride, were used to determine the selectivity of the biosensor. All three metals, especially cadmium and cobalt, at high concentration were found to elicit a response in the whole-cell biosensor; however, such response was much lower than that to zinc ions. The actual levels of cadmium in vivo are extremely low, except in the case of accidental exposure; therefore, cadmium is not expected to interfere with zinc sensing. On the other hand, cobalt is an essential metal that forms complexes with vitamin B_{12}; however, the physiological levels of cobalt in blood are 3 orders of magnitude lower than the physiological levels of zinc, thus also not interfering with zinc sensing.

10.3 Application of Spore-Based Whole-Cell Biosensors to Physiological and Environmental Samples

Although the previously described work demonstrated the use of spores as stable vehicles for biosensors, it did not investigate the application of these sensors to actual samples of physiological or environmental origin. It is very important when creating a sensor to ensure that the sensor still maintains its analytical characteristics when in the presence of sample matrix. To this end, Date et al. used the previously described biosensors for arsenic and zinc in both blood serum and freshwater [24]. The cells were sporulated using the exhaustion method described previously in this book chapter. After sporulation, 1.5 mL of the spore suspension was placed in 1.5 mL microcentrifuge tubes and centrifuged for 1 min at 10,000 rpm. The supernatant was then discarded and the spores were stored at room temperature until needed for the studies.

Spores have to germinate to vegetative cells for use as whole-cell biosensors, and the germination time needs to be optimized to achieve an assay that is as rapid as possible, while maintaining the analytical characteristics of the sensor. To that end, germination studies were performed to determine the time required for the spores to return to a vegetative state when placed in a nutrient-rich medium. Specifically, 750 μL of LB broth was added to each of the microcentrifuge tubes containing the arsenic sensing spores to trigger germination. To estimate the number of spores in each suspension, the OD_{600} of triplicate 45 μL aliquots was measured on a microtiter plate reader, which resulted in OD_{600} values of 0.60–0.65. This is an important step to ensure that the number of spores in each sample does not significantly differ so that the signal changes in an assay are actually due to changes in the response of the sensor and not to differences in cell concentration. Standard solutions of sodium arsenite (1×10^{-3}–1×10^{-9} M) were prepared in reverse-osmosis water by serial dilution of a 1×10^{-2} M stock solution in reverse-osmosis water. LB broth spore aliquots of 45 μL were added to the wells of a microtiter plate and allowed to incubate for 1, 2, or 3 h at 37°C and 250 rpm for germination. At each time point, 5 μL of the arsenite standards or reverse-osmosis water as blanks were added in triplicate to the germinating spores/vegetative cells and were incubated for 1 h at 37°C and 250 rpm. Then, 50 μL of the chemiluminescent substrate was added to each well and allowed to incubate for 30 min at 37°C, 250 rpm, after which time the luminescence signal was measured at 540 nm on a Polarstar Optima luminometer. The results of the assay demonstrated that 2 h is the minimum time for the spores to germinate and generate a sufficient amount of vegetative cells able to respond to the analyte and get a detection limit for arsenite of 1×10^{-7} M and a dynamic range of 1×10^{-4}–1×10^{-7} M. Subsequently, spores were directly incubated with arsenite standard solutions, allowing germination to occur in the presence of the analyte. An overall incubation time of 2 h, including spore germination and expression of the reporter protein, was sufficient to achieve the same detection limit and dynamic range. Therefore, this incubation time and setup was used for all the next experiments with the arsenic spore-based sensing system.

The spore-based whole-cell sensing system was then evaluated as a viable method for detection of analytes in real samples, such as freshwater and blood serum. Freshwater was simulated using 384 mg/mL $NaHCO_3$, 240 mg/mL $CaSO_4 \cdot H_2O$, 240 mg/mL $MgSO_4$, and 16 mg/mL KCl in reverse-osmosis water, pH 7 [25]. Arsenite solutions ranging from 1×10^{-3} M to 1×10^{-9} M were prepared in simulated freshwater or commercially available human serum by serial dilution of a 1×10^{-2} M stock solution of sodium arsenite in reverse-osmosis water. In serum, a detection limit of 1×10^{-7} M arsenite and a dynamic range of 1×10^{-4}–1×10^{-7} M were found. The presence of serum decreased the chemiluminescence signal intensities as compared to the assay performed with arsenite standard solutions in reverse-osmosis water. This matrix effect may be due to proteins that are present in the serum; these proteins may interact with the analytes, thus making them less accessible to the whole-cell biosensor, which may cause a decrease in signal intensity. In the present assay, despite such signal decrease, the detection limits did not seem to be affected. Freshwater, on the other hand, did not affect either the signal intensity, or linear range, or the detection limit of the assay, which was 1×10^{-7} M. It is important to point out that this value is below the arsenic standard of 10 ppb in drinking water set by the US Environmental Protection Agency (EPA). Upon evaluating laboratory-spiked samples, 23 water samples from various sources, including tap water from city and private water systems, and natural surface water from lakes, ponds, rivers, and creeks, were assayed for arsenite. To validate the whole-cell biosensor, the tap and surface water samples were tested by the graphite furnace atomic absorption spectrometry (GFAAS) method, which has a detection limit of 5 ppb. No detectable arsenite was found in the tap and surface waters and these results were confirmed by GFAAS. In general, when developing a new analytical method, it is important to validate it with an established reference method to ensure the new method is sensitive and accurate enough to be competitive with existing methods.

In conclusion, these studies demonstrated that the spores can successfully be used as vehicles for whole-cell biosensors and be germinated in the presence of complex matrixes, such as freshwater and serum. Importantly, there is no loss in the analytical performance of the sensor in the tested sample matrixes for both the described pMUTin-23-based sensor for arsenic, which employs β-galactosidase as the reporter, and the aforementioned pSD202-based sensor for zinc, which utilizes GFP as the reporter protein. Additionally, the sensor can convert from spores to germinated cells and allow for sensing and detection in an efficient time of less than 2.5 h.

10.4 Extreme Conditions

Although whole-cell biosensors have been employed for years, few reports discuss the use of these sensors in the field under extreme or harsh conditions. When aiming at field applications, it is important to consider the environment in which the sensor will be used and the stability of that sensor under the specified environmental conditions. Biosensors packaged in spores are ideal candidates for use in harsh environments due to the spore innate stability. Sangal et al. demonstrated the suitability of the spore as a packaging tool for whole-cell sensing systems under various laboratory conditions resembling those of natural extreme environments, such as desiccation to represent a dry climate with an average temperature of 25°C, a water bath at a temperature of 28°C to simulate wet heat as in a tropical area, an oven set at 37°C to simulate dry heat seen in desert settings, and a freezer at –20°C to simulate the cold found in arctic and polar areas [26]. To examine the effects of these simulated environments on spore-based whole-cell biosensors, *B. subtilis* cells harboring the plasmid pMUTin-23 were employed as a model. This whole-cell sensing system had already been shown to be a viable and robust sensing system for arsenic after months of storage at room temperature in the form of spores. In the study described herein, the analytical performance of the sensing system was evaluated to define the reference analytical characteristics for comparison upon storage in the above conditions. Although the system was previously characterized, it is always necessary to confirm the reported features when proposing new applications, since different cell batches, reagent lots, and instruments may be employed. Initially, a single colony of *B. subtilis* cells harboring the plasmid pMUTin-23 was inoculated into 300 mL of LB broth and grown at 37°C and 250 rpm. When the cells reached an OD_{600} of 0.6–0.7, 50 μL aliquots were transferred to the wells of a 96-well microtiter plate along with triplicate 10 μL aliquots of standard solutions of potassium antimonyl tartrate, sodium arsenite, and sodium arsenate of concentrations of 1×10^{-3} to 1×10^{-9} M, which were prepared by serial dilution of a 1×10^{-2} M stock solution in reverse-osmosis water. Additionally, 10 μL of reverse-osmosis water was added in triplicate to the cells as a blank. This resulted in final analyte concentrations of 1.7×10^{-4} to 1.7×10^{-10} M. The cells and analytes were then incubated for 1 h at 37°C, 250 rpm. After incubation, 60 μL of Beta-glo reagent was added to each well and incubated for 30 min until readings were taken on a Polarstar Optima microtiter plate reader with an integration time of 3 s per well. The amount of β-galactosidase produced by the whole-cell biosensor was directly proportional to the amount of analyte present. It was found that at the highest concentration of analyte, the signal decreased. This was due to the toxicity of the analyte on the cells at that concentration. At analyte concentrations below the highest, the sensor produced a dose-dependent response for arsenite, arsenate, and antimonite with detection limits of 1.7 $\times 10^{-7}$, 5×10^{-8}, and 5×10^{-8} M, respectively. Additionally, the percentage relative standard deviation was calculated for the assay at values less than 10%, indicating that the assay was reproducible. One thing to note about the use of the 6-O-β-galactopyranosyl-luciferin/luciferase system contained in the Beta-glo reagent was that the signal intensity was so strong that the differences between analyte concentrations were visible by the naked eye, with the higher concentrations having a more intense color that decreased as the analyte concentration decreased (Fig. 10.6).

After the assay was performed with fresh vegetative cells, spores were generated, as previously described in this book chapter. Briefly, cells were grown in a sporulation medium at 37°C, 250 rpm, for 4 days, and then the obtained spores were centrifuged at 10,000x

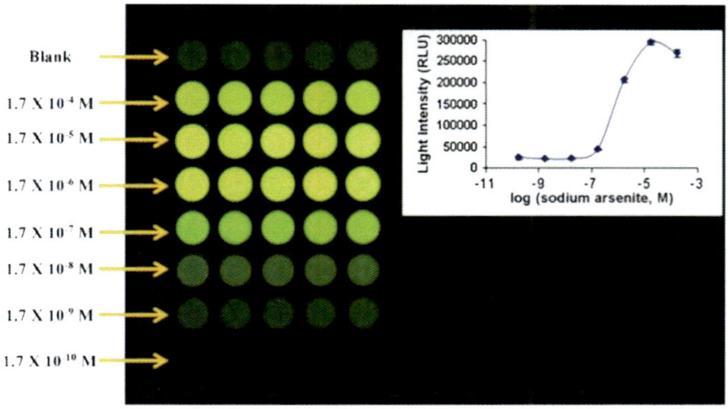

Figure 10.6 Microtiter plate showing the light intensity change with decreasing arsenite concentrations. Although detection with a microtiter plate reader is more sensitive and quantifiable, this method allows for analyte concentration differentiation with the naked eye. Adapted with permission from Ref. [26]. Copyright (2011) Elsevier.

g for 10 min at 4°C. The resulting pellet was washed twice and resuspended in reverse-osmosis water, sonicated for five min, and then centrifuged at 10,000x g for 10 min at 4°C to release the spores from the cell debris. The pellet was finally resuspended in 120 mL of reverse-osmosis water, which was then transferred to 72 microcentrifuge tubes in 1.5 mL aliquots. These tubes were centrifuged at 10,000x g for 10 min at room temperature, the supernatants were discarded, and the spores were stored in the various simulated environmental conditions described above, namely dry heat, wet heat, desiccation, and cold for 12 months. Each month, one tube from each environmental condition was taken and the spores resuspended in 750 μL of LB broth to begin germination. The suspension was then inoculated into a 300 mL culture of LB broth and grown at 37°C, 250 rpm until the OD_{600} reached 0.6–0.7, which took about 15 h. The cultures were then evaluated for their ability to sense the three analytes: arsenite, arsenate, and antimonite. After 12 months, the sensor was found to have a detection limit of 2.0 $\times 10^{-8}$ M for arsenite under conditions of wet heat, dry heat, and desiccation. At cold temperature, the detection limit for arsenite was 5.0×10^{-8} M. All of these values were similar to that of the fresh vegetative cells, which had a detection limit of 1.7×10^{-7} M for arsenite. The germinated cells had a detection limit of 5×10^{-8} M for antimonite in all four storage conditions, which was equivalent to the detection limit of the fresh vegetative cells for antimonite. Arsenate showed a detection limit of 5×10^{-8} M in dry heat and desiccation and 5×10^{-7} M in cold and wet heat, all of which are similar to the detection limit of 5×10^{-8} M of the fresh vegetative cells. Additionally, the dynamic range of the germinated cells at each time point spanned 2 orders of magnitude, the same as the vegetative cells before sporulation and storage. The assay was also shown to be reproducible over time with relative standard deviation (%RSD) values less than 9%. It is important to note that lower detection limits were found during the seventh and eighth months of the study. The authors hypothesized that this may be due to the use of different batches of the luminescence detection kit. This demonstrates that it is important to complete all experiments with the same batch of reagents as the quality and activity of certain reagents may vary between batches. Additionally, the change in detection limit at certain time points may be due to the occasional variability intrinsic to living cells. Although the optical density of the starting spore suspensions can be normalized between assays, it is difficult to control the response and behavior of each batch of sensing cells.

This study demonstrated that the robustness of the spore makes it an ideal candidate for use as a vehicle for whole-cell biosensors, not only in mild conditions, but also in extreme ones. The spores can be stored under various environmental conditions, including dry heat, wet heat, cold, and desiccation, be germinated, and retain the sensing characteristics of the fresh vegetative cells for at least 1 year. This is particularly important for the successful use of biosensors in developing countries and remote areas, where harsh environments and poor storage facilities are often found.

10.5 Packaging of Spores for Field Applications

As described above, spores have proven to be stable vehicles for whole-cell biosensors. They are able to germinate and generate viable sensing cells with unchanged analytical characteristics upon storage for at least 2 years at room temperature and 1 year under extreme environmental conditions. This makes them ideal for field applications; however, to be used in the field, portable, easy-to-use, and cost-effective platforms must be developed that can serve for both storage and transportation of the spores and for germination and analyte detection, when needed. To this end, several methods based on microfluidics and paper strip supports have been developed and proposed as tools for field analysis.

10.5.1 *Microfluidic Platforms*

Microfluidic total analysis systems (μTAS) are an attractive means for diagnostic and other high-throughput assays. μTAS provide affordable systems by decreasing assay times and the consumption of reagents and samples, reducing device size, and having relatively easy operation. These systems are able to combine multiple analytical functions such as mixing, sorting, and trapping to perform analysis on a single platform. They have been used for a variety of applications, including cell lysis and DNA purification [27], real-time quantitative PCR [28], and others. Furthermore, these devices can be made of glass, silicon, or inexpensive polymer materials such as poly(dimethyl siloxane) (PDMS) and poly(methyl methycrylate) (PMMA) [29]. The design of the microfluidic channels and reservoirs can be manufactured by several wet or dry methods, including photolithography, etching, and injection molding, to include a few [30]. Electrokinetic pumping and centrifugal pumping are some of the most common means to control propulsion of the fluid through the device. Electrokinetic pumping works by generating a charged double layer that produces a bulk fluid flow when a potential is applied to the system; however, this method would be damaging to live cells that are used in whole-cell biosensors due to the need for a high voltage [31, 32]. Therefore, centrifugal pumping is employed when live cells are used. In this method, μ-dimension architectures are machined onto disk-shaped polymer substrates, which are similar in dimension to a compact disk (CD) [33]. The fluid flow is dependent on the angular frequency of rotation and channel dimensions of the CD platform and on the physical properties of the fluid. Cell-based and protein-based sensors have previously been adapted to centrifugal CD microfluidic devices [34–36]. There are numerous detection methods that can be used on these platforms, including electrochemical, optical, and mass spectrometry.

The fabrication of the CD microfluidic device has been previously described [37]. The structure of the CD was designed using a computer-aided design (CAD) program. For each application, the number of reservoirs and mixing channels can be changed to suit the experiment. The design was then computer numerical control (CNC)-machined onto a PMMA substrate that was 3 mm thick. Solutions were then pipetted into the reservoirs and each structure was covered with an optically transparent tape. To reduce background pressure created when the fluids move to the edge of the structure during spinning, air vents were punctured with a syringe.

In the example of arsenic sensing, the CD had six identical structures and the template measured 120 mm in diameter with a hole in the center that was 15 mm in diameter. A burst valve was fabricated for chamber 1, which measured 635 μm in width and depth. Due to the fact that this CD was designed to reconstitute the spores, chamber 1 would burst at a lower angular frequency than the subsequent chambers and release the reagent into reservoir 2, which would harbor the spores. Each structure consisted of two reagent reservoirs, 2 and 3. These reservoirs were both 127 μm wide and 63.5 μm in depth. Since these reservoirs were equidistant from the center of the disk, the centrifugal force required for release was equivalent for both of these chambers. The analyte solution/sample was kept in chamber 3 and when the centrifugal force was high enough to overcome surface tension, both of these valves burst. Reservoirs 2 and 3 were connected via channels to a mixing channel, which terminated in the detection chamber, 4 (Fig. 10.7b). The mixing channel was designed to be a serpentine structure to facilitate mixing of the analyte and the reporter cells. This structure was important because the flow would normally be laminar at these channel dimensions; however, the sharp 90° angles of the serpentine structure allowed for tumbling of the fluids, which resulted in a more thorough mixing.

For the measurement setup, a custom system was built. The disk was fit into a motor hub and secured. This hub was run via a controller (PIC-Servo, HdB Electronics, Redwood City, CA). This controller was powered by a 12 V battery and interfaced to a PC by a Visual Basic program. The software was designed to allow for the rotating platform to accelerate to specific angular frequencies for the burst valves. After the valves were burst and the reagents were mixed, the CD was decelerated to allow for the fluorescence measurements. The measurements were performed by manually rotating the disk and reading on a Varian Cary Eclipse fluorescence spectrometer, which had a fiber optic attachment. The fiber optic was placed perpendicular to the CD at a distance of 2 mm above the center of the detection chamber.

The CD structure shown in Fig. 10.7a was used for zinc sensing. This structure is similar to that of Fig. 10.7b but has one less reagent chamber. The extra chamber was not needed, because the reporter in the zinc whole-cell biosensing system was GFP and did not call for the addition of a substrate, unlike the arsenic biosensor that did need a substrate. This demonstrates the versatility of the design of these microfluidic platforms. For this reason, CD microfluidic

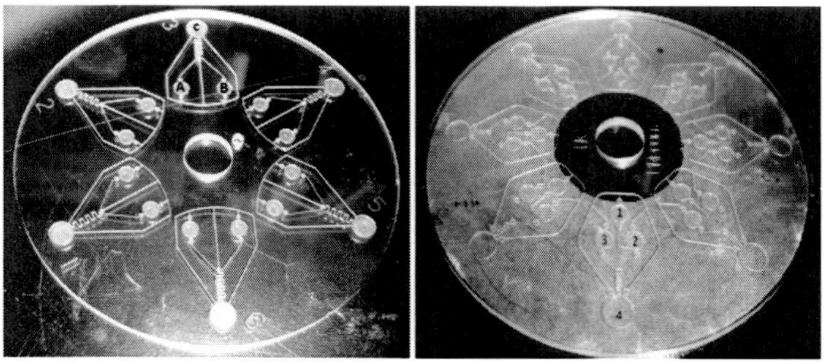

(a) (b)

Figure 10.7 (a) A CD microfluidic platform was manufactured with six identical structures for the detection of Zn^{2+}. Each structure has two reagent reservoirs (A and B), one for analyte standard solutions or samples and one for spore suspensions, and a detection chamber (C), which are connected by a serpentine mixing channel. (b) For arsenic detection, a CD microfluidic device with eight identical structures was manufactured. Reservoir 1 was employed for analyte standard solutions or environmental and physiological samples, reservoir 2 for spore suspensions, reservoir 3 for the chemiluminescent substrate, and reservoir 4 for detection. Adapted with permission from Ref. [36]. Copyright (2010) Springer-Verlag.

devices are ideal platforms for the storage of spores as vehicles for biosensing systems and for the detection of analytes using the spore-based whole-cell biosensors.

The *B. megaterium* and *B. subtilis* spore-based whole-cell sensing systems for the detection of zinc and arsenic, as described earlier, in this chapter were employed to demonstrate the use of the CD microfluidic devices for storage, assay, and detection of analytes [38]. For the zinc sensor, a 120 cm diameter circular disk made of PMMA was employed. Initially, a time study was performed on the CD to determine the optimum incubation time of the spores in a nutrient-rich medium for germination. To do this, 17 μL of a *B. megaterium* spore suspension in LB broth at an OD_{600} of 0.35–0.40 was loaded into reagent reservoir A and 17 μL of a 2×10^{-4} M zinc chloride solution in reverse-osmosis water, or reverse-osmosis water as a blank, was loaded into reservoir B. The disk was then spun at 950 rpm for 30 s. This speed was high enough to reach the burst frequency and the spores and analyte were mixed in the mixing channel before reaching the detection chamber, C. Fluorescence measurements were taken at 20 min intervals beginning at 10 min after the mixing for up to 210 min. The data indicated that the spores transformed into metabolically active cells that could respond to the presence of the analyte in 90 min and the fluorescence intensity continued to rise until 210 min.

Additionally, a time study was performed by incubating the spores with lower concentrations of the zinc standard solutions. The lowest zinc concentration required a minimum incubation time of 150 min before detection was possible. Therefore, the incubation time of 150 min was chosen to achieve a dose–response curve. A detection limit of 1×10^{-6} M and a dynamic range of 1×10^{-6} to 1×10^{-4} M were obtained, which were comparable with those found when the assay was performed on a microtiter plate. The inter- and intra-assay reproducibility showed CVs below 10%. The microfluidic format assay was also performed in serum samples spiked with known concentrations of zinc.

A dose–response curve was generated, as described above, and a detection limit of 1×10^{-6} M with a dynamic range of 1×10^{-6} to 1×10^{-4} M was obtained in serum. The fluorescence intensity in the serum was higher than the fluorescence in reverse-osmosis water.

This may be due to the fluorescence background from proteins in serum. Additionally, zinc is known to be present in serum from healthy subjects at concentrations between 12 and 14 μM, which may have also contributed to the fluorescence signal increase in the serum.

For the arsenic sensor, a 12 cm diameter PMMA CD with eight identical structures was employed (Fig. 10.7b). There are four chambers in each structure on the CD. To perform the assay, 8 μL of the sodium arsenite standard solutions ranging from 2×10^{-3} to 2×10^{-8} M prepared in reverse-osmosis water, or reverse-osmosis water as a blank, was placed in reservoir 1. To reservoir 2, 8 μL of spores suspended in LB broth was added. Finally, 8 μL of the chemiluminescent substrate for β-galactosidase detection was placed in reservoir 3. Reservoir 1 has a burst frequency that differs from that of reservoirs 2 and 3, which instead have equal burst frequencies. Therefore, the CD was initially spun at 700 rpm for 30 s to allow the analyte solution from reservoir 1 to be released and mix with the spore suspension in reservoir 2. The CD was then incubated at room temperature for 150 min for the germination of the spores and production of the reporter protein to occur. After this incubation, the CD was spun at 1000 rpm for 30 s to release the vegetative cells incubated with analyte and the chemiluminescent substrate from reservoir 2 and reservoir 3, respectively, to the detection chamber, 4. Measurements were taken after a 30 min incubation time and a dose–response curve was generated. A detection limit of 1×10^{-7} M arsenite with a dynamic range of 1×10^{-7} to 1×10^{-4} M and inter- and intra-assay CVs below 10% were found on the CD, which is consistent with previous results that were obtained in a microtiter plate. The arsenic assay was also performed with serum and freshwater to determine if the sample matrix affects the analytical performance of the spore-based sensing system in the miniaturized CD format. Similarly to the characterization of the zinc sensor, dose–response curves were generated by spiking the test matrixes with known concentrations of arsenite. Again, a detection limit of 1×10^{-7} M of arsenite in both freshwater and serum was found with a dynamic range of 1×10^{-7} to 1×10^{-4} M, indicating that no significant matrix effects were observed; however, similarly to the assay performed in microtiter plate, there was a decrease in the chemiluminescence signals in the spiked serum samples as compared to the standards in reverse-osmosis water. The authors hypothesized that the decrease was due to the arsenite binding to proteins in the serum, thus being less available for diffusion into the sensing cells. This points out to the need of generating a dose–response curve in the proper matrix to be employed as a reference when detecting analytes in various environmental and physiological matrixes in order to perform accurate quantitative analysis.

In summary, by utilizing a compact disk microfluidic platform, the analysis of samples was performed in less than 3 h, including the time for germination and detection, while employing small (low microliter range) volumes of reagents and samples. The authors have shown that these CD microfluidic platforms are μTAS in which all chemical and biological reagents can be stored and all analytical steps can be performed on a single miniaturized device, which makes them portable analytical systems suitable for onsite applications.

10.5.2 *Paper-Based Sensors*

Although the CD microfluidic platform described above does enable onsite analysis, there is still a need for relatively expensive equipment to detect and quantify the amount of the reporter produced and, consequently, of the target analyte present. To provide a simpler, inexpensive, and portable assay not requiring instrumentation, Fantino et al. proposed the development of a so-called sposensor, which uses a colorimetric detection method that, for initial analysis, can be viewed with the naked eye for an estimation of the analyte present [39]. The authors employed a fusion of the *czcD* gene to the *lacZ* reporter gene, which was integrated at the *amyE* locus on the chromosome of the HB5016 *B. subtilis* strain. The *cdcZ* gene is induced in the presence of metals such as Zn^{2+}, Ni^{2+}, Cd^{2+}, and Cu^{2+}, thus increasing the expression of β-galactosidase reporter [40]. Initially, the HB5016 spores were grown on a solid medium, supplemented with 5-bromo-4-chloro-3-indoyl-β-D-galactopyranoside chromogen at a 0.1 mg/mL final concentration, containing Zn^{2+}, Ni^{2+}, Cd^{2+}, or Cu^{2+}. The β-galactosidase activity was monitored by development of blue color after overnight incubation at 30°C. Unexpectedly, the cells only showed a response to Zn^{2+}, indicating that this cell strain was specific to Zn^{2+} detection in the employed conditions. Therefore, the rest of the studies were focused on only Zn^{2+} as the analyte. Next, vegetative HB5016 *B. subtilis* cells were grown in a liquid growth medium with various concentrations of Zn^{2+}. A dose-dependent, linear relationship was found for the *cdcZ* expression and Zn^{2+} concentration ranging from 10 to 300 μM. Since the system specifically responded to Zn^{2+} in a dose-dependent manner, the authors proceeded to the development of the sposensor. The system was designed to be a three-tube kit with a one-step incubation/detection process: the first tube contains spores immobilized on paper disks, the second tube contains a nutrient medium to begin the germination of the spores, and the third tube contains a bactericidal solution (mix of $CuCl_2$ and ascorbic acid) to be added to the assay tube after the completion of the measurement. Spores were made by incubation of the *B. subtilis* cells in Schaeffer's sporulation medium (SSM) for 48 h. The spores were then harvested by centrifugation at 15,000 g for 15 min and resuspended in 0.1 M Tris-KCl buffer at pH 8 to an OD_{600} of 2. The suspension was then treated at 80°C for 10 min, after which time an equal volume of prewarmed Tris-KCl buffer containing 4% agar was added, and 20 μL aliquots of the mixture were spotted on autoclaved Whatman 3 filter paper disks. In addition to the sensing spores, control spores were prepared from cells of the *B. subtilis* SG57 strain that constitutively expresses a reporter to monitor spore germination and cell viability. Disks containing either the sensing or control spores were finally placed in the cap of the first tube. The second tube contains a dried powder medium that is composed of 100 mg bacto-tryptone, 50 mg bacto-yeast extract, 10 mg NaCl, 20 mg glucose, and 1 mg 5-bromo-4-chloro-3-indolyl-β-D-galactopyranoside and is reconstituted with 2 mL of analyte solution to be tested in distilled water. The assay was performed by transferring the paper disks assembled in the cap of the first tube to the top of the second tube, which was then incubated in a horizontal position so that the disks were kept at the surface of the solution to avoid spore dissemination and allow for germination of the spores on the disks. After an incubation time of 16 h in a horizontal position at 24°C, 30°C,

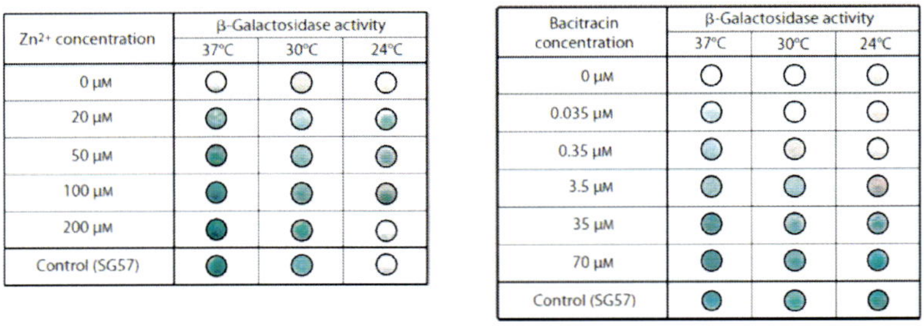

(a) (b)

Figure 10.8 (a) Color development of the sposensor response to Zn^{2+}. (b) Color development of the sposensor response to bacitracin. Adapted with permission from Ref. [37]. Copyright (2009) S. Karger AG, Basel.

and 37°C, 2 mL of the bactericidal solution was added to destroy vegetative cells and ungerminated spores. After incubating with various concentrations of zinc, a dose-dependent difference in blue color intensity could be identified on the filter paper for all three temperatures; however, at 24°C, the difference in coloration could only be detected for concentrations of zinc lower than 100 μM as higher concentrations did not yield any color (Fig. 10.8a). The authors hypothesized that high concentrations appeared to be toxic to the cells at that temperature since the control cells did not show any color development either when incubated in the presence of a zinc concentration higher than 100 μM. The assay had a detection limit of 10 μM Zn^{2+} at all three tested temperatures, which is almost eight times lower than the allowed US EPA maximum drinking water concentration of Zn^{2+} and equivalent to the detection limit achieved by the vegetative cells, indicating that sporulation and germination did not affect the analytical characteristics of the sensor. To evaluate the versatility of the sposensor approach for sensing of analytes other than metals, the authors chose to detect bacitracin, which is a peptide antibiotic. A plasmid was constructed with a *bceA* gene, which is part of a group of genes encoding an ABC transporter that allows for *B. subtilis* cells to respond and resist to bacitracin, fused to the *lacZ* gene. When bacitracin is present in the medium, it induces the transcription of the *bceA* gene, which in turn begins the transcription of *lacZ*. The assay was performed employing spores in the three tube format, as described previously. The data showed a detection limit of 0.035 μM of bacitracin at 37°C (Fig. 10.8b). To confirm these results, the authors also used vegetative cells that were grown from 1 μL spore suspension overnight and subcultured to adjust the OD_{600} to 0.05 and then grown at 37°C with rapid aeration until the OD_{600} reached 0.4–0.6. The cells were incubated with various concentrations of bacitracin for 30 min at 37°C and then rapidly chilled on ice. β-galactosidase activity was determined using *o*-nitrophenyl-β-galactopyranoside as the substrate [41]. It was found that with the vegetative cells the detection limit was lower than 70 nM of bacitracin, with the linear response between 70 nM and 7 μM.

In conclusion, the sposensor has an advantage over the microfluidic device for field applications, namely there is no need for expensive instrumentation, because the dose-dependent color change of the sposensor can be observed by the naked eye. These paper strip sensors have been used previously for the detection of analytes that are biologically relevant [42]. Further information on the preparation of paper-incorporated sensing cells can be found in Ref. [41]. The microfluidic device, however, provides a more rapid detection system with results obtained in 150 min, including spore germination and sensing, compared to the sposensor, which requires 16 h of incubation. Additionally, the CD microfluidic sensor can effectively be operated at room temperature, while the sposensor needs controlled temperature for optimal performance.

10.6 Conclusions

When developing portable sensors for onsite applications, it is important to achieve a detection system that does not require expensive instrumentation, is user friendly, has a minimal assay time, and has adequate stability under various conditions. The methods described in this chapter have demonstrated the stability of spores as vehicles for preservation, storage, and transport of whole-cell-based biosensors under various environmental conditions. Specifically, the long-term stability of spore-based whole-cell sensing systems, showing a shelf life of over 2 years at room temperature and at least 1 year in harsh conditions, suggests that the development of whole-cell biosensors based on spore-forming bacteria can be a successful strategy to increase the robustness of cell-based biosensing systems and expand their onsite applications to a variety of locations, including remote areas and developing countries, which often face challenging climates and insufficient facilities. Various detection techniques ranging from fluorescence, bioluminescence, and chemiluminescence to colorimetric changes have been proven to be successful when employing spore-based whole-cell sensing systems, thus providing a variety of measurement approaches that are amenable to field applications. This demonstrates that spores as vehicles for whole-cell biosensors are only limited by the biosensor capabilities; sporulation and subsequent germination do not interfere with the analytical characteristics of the whole-cell biosensor over long-term storage in various settings.

Acknowledgments

The authors would like to thank the National Science Foundation (grant CHE-0718844) and the National Institutes of Environmental Health Sciences (grant P42ES07380) for partly funding this work. SD thanks the Miller School of Medicine of the University of Miami for the Lucille P. Markey Chair in Biochemistry and Molecular Biology, as well as the University of Kentucky for a Gill Eminent Professorship. LDK acknowledges a National Science Foundation Integrative Graduate Education Research Traineeship for funding. The authors would also like to thank Trajen Head for his assistance with the figures for this work.

References

1. Cano, R. J., Borucki, M. K. (1995). Revival and identification of bacterial spores in 25- to 40-million-year-old Dominican amber. *Science*, **268**(5213), 1060–1064.
2. Van Mellaert, L., Barbe, S., Anne, J. (2006). Clostridium spores as anti-tumour agents. *Trends Microbiol.*, **14**(4), 190–196.

3. Barbe, S., Van Mellaert, L., Anne, J. (2006). The use of clostridial spores for cancer treatment. *J. Appl. Microbiol.*, **101**(3), 571–578.
4. Errington, J. (2010). From spores to antibiotics via the cell cycle. *Microbiology*, **156**(Pt 1), 1–13.
5. Ricca, E., Cutting, S. M. (2003). Emerging applications of bacterial spores in nanobiotechnology. *J. Nanobiotechnol.*, **1**(1), 6.
6. Driks, A. (2003). The dynamic spore. *Proc. Natl. Acad. Sci. U S A*, **100**(6), 3007–3009.
7. Atrih, A., Foster, S. J. (2002). Bacterial endospores the ultimate survivors. *Int. Dairy J.*, **12**(2–3), 217–223.
8. Setlow, P. (2001). Resistance of spores of Bacillus species to ultraviolet light. *Environ. Mol. Mutagen.*, **38**(2–3), 97–104.
9. Setlow, P. (2006) Spores of Bacillus subtilis: their resistance to and killing by radiation, heat and chemicals. *J. Appl. Microbiol.*, **101**(3), 514–525.
10. Setlow, P. (1992). I will survive: protecting and repairing spore DNA. *J. Bacteriol.*, **174**(9), 2737–2741.
11. Yi, X., Setlow, P. (2010). Studies of the commitment step in the germination of spores of Bacillus species. *J. Bacteriol.*, **192**(13), 3424–3433.
12. Kumari, A., Pasini, P., Deo, S. K., Flomenhoft, D., Shashidhar, H., Daunert, S. (2006). Biosensing systems for the detection of bacterial quorum signaling molecules. *Anal. Chem.*, **78**(22), 7603–7609.
13. Michelini, E., Guardigli, M., Magliulo, M., Mirasoli, M., Roda, A., Simoni, P., Baraldini, M. (2006). Bioluminescent biosensors based on genetically engineered living cells in environmental and food analysis. *Anal. Lett.*, **39**(8), 1503–1515.
14. Gu, M. B., Mitchell, R. J., Kim, B. C. (2004). Whole-cell-based biosensors for environmental biomonitoring and application. *Adv. Biochem. Eng. Biotechnol.*, **87**, 269–305.
15. Abernathy, C. O., Liu, Y. P., Longfellow, D., et al. (1999). Arsenic: health effects, mechanisms of actions, and research issues. *Environ. Health Perspect.*, **107**(7), 593–597.
16. Date, A., Pasini, P., Daunert, S. (2007). Construction of spores for portable bacterial whole-cell biosensing systems. *Anal. Chem.*, **79**(24), 9391–9397.
17. Erbe, J. L., Taylor, K. B., Hall, L. M. (1995). Metalloregulation of the cyanobacterial Smt locus: identification of Smtb binding sites and direct interaction with metals. *Nucleic Acids Res.*, **23**(13), 2472–2478.

18. Sato, T., Kobayashi, Y. (1998). The ars operon in the skin element of Bacillus subtilis confers resistance to arsenate and arsenite. *J. Bacteriol.*, **180**(7), 1655–1661.
19. Harwood, C. R., Cutting, S. M. (1990). *Molecular Biological Methods for Bacillus*, Wiley, New York, USA.
20. Date, A., Pasini, P., Daunert, S. (2010). Fluorescent and bioluminescent cell-based sensors: strategies for their preservation. *Adv. Biochem. Eng./Biotechnol.*, **117**, 57–75.
21. Black, M. M. (2003). The evidence linking zinc deficiency with children's cognitive and motor functioning. *J. Nutr.*, **133**(5 Suppl 1), 1473S–1476S.
22. Vary, P. S. (1994). Prime time for Bacillus megaterium. *Microbiology*, **140** (Pt 5), 1001–1013.
23. Turner, J. S., Glands, P. D., Samson, A. C., Robinson, N. J. (1996). Zn2+-sensing by the cyanobacterial metallothionein repressor smtb: different motifs mediate metal-induced protein-DNA dissociation. *Nucleic Acids Res.*, **24**(19), 3714–3721.
24. Date, A., Pasini, P., Sangal, A., Daunert, S. (2010). Packaging sensing cells in spores for long-term preservation of sensors: a tool for biomedical and environmental analysis. *Anal. Chem.*, **82**(14), 6098–6103.
25. Griffin, R. A., Jurinak, J. J. (1973). Estimation of activity-coefficients from electrical conductivity of natural aquatic systems and soil extracts. *Soil Sci.*, **116**(1), 26–30.
26. Sangal, A., Pasini, P., Daunert, S. (2011). Stability of spore-based biosensing systems under extreme conditions. *Sens. Actuators B*, **158**(1), 377–382.
27. Ha, S. M., Cho, W., Ahn, Y., Hwang, S. Y. (2011). An integrated microfluidic device for rapid cell lysis and DNA purification of epithelial cell samples. *J. Nanosci. Nanotechnol.*, **11**(5), 4250–4253.
28. White, A. K., VanInsberghe, M., Petriv, O. I., Hamidi, M., Sikorski, D., Marra, M. A., Piret, J., Aparicio, S., Hansen, C. L. (2011). High-throughput microfluidic single-cell RT-qPCR. *Proc. Natl. Acad. Sci. U S A*, **108**(34), 13999–14004.
29. Friend, J., Yeo, L. (2010). Fabrication of microfluidic devices using polydimethylsiloxane. *Biomicrofluidics*, **4**(2), 026502.
30. Mijatovic, D., Eijkel, J. C. T., van den Berg, A. (2005). Technologies for nanofluidic systems: top-down vs. bottom-up: a review. *Lab Chip*, **5**(5), 492–500.

31. Soper, S. A., Ford, S. M., Qi, S., McCarley, R. L., Kelly, K., Murphy, M. C. (2000). Polymeric microelectromechanical systems. *Anal. Chem.*, **72**(19), 642a–651a.

32. Madou, M., Zoval, J., Jia, G., Kido, H., Kim, J., Kim, N. (2006). Lab on a CD. *Annu. Rev. Biomed. Eng.*, **8**, 601–628.

33. Moschou, E. A., Nicholson, A. D., Jia, G. Y., Zoval, J. V., Madou, M. J., Bachas, L. G., Daunert, S. (2006). Integration of microcolumns and microfluidic fractionators on multitasking centrifugal microfluidic platforms for the analysis of biomolecules. *Anal. Bioanal. Chem.*, **385**(3), 596–605.

34. Cooney, C. G., Sipes, D., Thakore, N., Holmberg, R., Belgrader, P. (2012). A plastic, disposable microfluidic flow cell for coupled on-chip pcr and microarray detection of infectious agents. *Biomed. Microdevices*, **14**(1), 45–53.

35. Yin, H., Marshall, D. (2012). Microfluidics for single cell analysis. *Curr. Opin. Biotechnol.*, **23**(1), 110–119.

36. Park, J., Sunkara, V., Kim, T. H., Hwang, H., Cho, Y. K. (2012). Lab-on-a-disc for fully integrated multiplex immunoassays. *Anal. Chem.*, **84**(5), 2133–2140.

37. Puckett, L. G., Dikici, E., Lai, S., Madou, M., Bachas, L. G., Daunert, S. (2004). Investigation into the applicability of the centrifugal microfluidics platform for the development of protein-ligand binding assays incorporating enhanced green fluorescent protein as a fluorescent reporter. *Anal. Chem.*, **76**(24), 7263–7268.

38. Date, A., Pasini, P., Daunert, S. (2010). Integration of spore-based genetically engineered whole-cell sensing systems into portable centrifugal microfluidic platforms. *Anal. Bioanal. Chem.*, **398**(1), 349–356.

39. Fantino, J. R., Barras, F., Denizot, F. (2009). Sposensor: a whole-bacterial biosensor that uses immobilized Bacillus subtilis spores and a one-step incubation/detection process. *J. Mol. Microbiol. Biotechnol.*, **17**(2), 90–95.

40. Moore, C. M., Gaballa, A., Hui, M., Ye, R. W., Helmann, J. D. (2005). Genetic and physiological responses of Bacillus subtilis to metal ion stress. *Mol. Microbiol.*, **57**(1), 27–40.

41. Miller, J. H. (1972). *Experiments in Molecular Genetics*, Cold Spring Harbor, NY, USA.

42. Struss, A., Pasini, P., Ensor, C. M., Raut, N., Daunert, S. (2010). Paper strip whole cell biosensors: a portable test for the semiquantitative detection of bacterial quorum signaling molecules. *Anal. Chem.*, **82**(11), 4457–4463.

Index